A Manager's Guide to

HEALTH
and
SAFETY
at
WORK

A Manager's Guide to

HEALTH
and
SAFETY
at
WORK

Jeremy Stranks

FIFTH EDITION

KOGAN PAGE

First published in 1990
Second edition published in 1992
Third edition published in 1994
This edition published in 1997

Kogan Page Limited
120 Pentonville Road
London N1 9JN

© Jeremy Stranks 1990, 1992, 1994, 1995, 1997

British Library Cataloguing in Publication Data
A CIP record for this book is available from the British Library.

ISBN 0 7494 2435 4

Typeset by J&L Composition Ltd, Filey, North Yorkshire
Printed and bound in Great Britain by Biddles Ltd, Guildford and Kings Lynn

Contents

List of Figures

List of Tables

Foreword

Managers must appreciate that the whole approach to health and safety at work has changed. No longer are we concerned with complying with the minimum prescriptive standards laid down in the old legislation. The need for managers to manage their responsibilities for health and safety is the key issue of the 1990s. But what do we mean by 'management'? One definition is: 'the effective use of resources in the pursuit of organisational goals'. People are undoubtedly an organisation's most important resource. Without people, we wouldn't be in business.

Under the Management of Health and Safety at Work Regulations 1992, organisations have a very important duty, namely, 'to make and give effect to such arrangements as are appropriate for the effective planning, organisation, control, monitoring and review of the preventive and protective measures'. This means incorporating health and safety procedures and systems in the normal day-to-day management of the business and not waiting until an accident occurs before taking action.

I am sure *A Manager's Guide to Health and Safety at Work* will go some way in providing the information to managers on the systems they must install and the measures they must take to ensure effective health and safety management.

Mr Frank Davies CBE
Chairman
Health & Safety Commission

Preface to First Edition

Health and safety at work is, for many managers, a difficult subject. Apart from being steeped in the law, which can be difficult to interpret, it requires a broad knowledge of many disciplines, such as psychology, engineering, chemistry, ergonomics and medicine, each of which is a subject of study in its own right.

Individual attitudes to health and safety, and indeed the corporate attitudes of organisations, may vary substantially. Is health and safety just a question of complying with the law? Should an organisation that likes to think of itself as caring, promote health and safety as part of that caring philosophy? Or should health and safety be seen as an important feature of the business operation aimed at reducing the losses associated with accidents, ill health, etc?

Increasingly, as we move towards a quality-orientated approach to our business activities, it will be seen that health and safety is an integral feature of such an approach. Everyone has a role to play in the management of quality. Accidents, ill health, sickness absence and damage-producing incidents feature in the 'price of non-conformance' – an important measure in quality management.

How much did accidents and sickness absence cost your organisation last year? What is the cost of your current employer's and occupier's liability insurance? Have you recently been prosecuted and fined for breaches of health and safety legislation? There is no doubt that accidents and sickness represent substantial losses to any company. A meagre 10 per cent reduction in these costs can be significant. A 20 per cent reduction would be marvellous.

There is a need, therefore, for managers to be more knowledgeable about the subject of health and safety from a legal, scientiôc and technical viewpoint. This book has been written with this objective in mind. I hope it will help managers to understand the subject better and make a contribution to improved levels of health and safety performance.

Jeremy Stranks
1989

Preface to Fifth Edition

Increasingly managers are having to become more knowledgeable about health and safety. This has been brought about as a result of the stringent duties under recent legislation, the need for risk assessment, the appointment of competent persons and the clear-cut duty on employers actually to manage their health and safety activities.

Since the fourth edition, new legislation has come into operation with respect to pregnant workers, young persons, hazardous substances, consultation with employees and construction work. In particular, the new Construction (Health, Safety and Welfare) Regulations impose important new requirements on all those involved in construction work of any description.

In preparing the fifth edition, I have paid special attention to a number of well-known courses directed at line managers, including the Institution of Occupational Safety and Health's 'Managing Safely' course. This course is proving particularly suitable as a means of instilling basic health and safety principles at middle management level. Many of these managers are the directors and senior managers of the future, and it is essential they have a good grasp of the legal principles and management systems in health and safety.

Jeremy Stranks
1997

Acknowledgement

The author wishes to place on record his thanks to the British Standards Institution.

Extracts from British Standards are reproduced by permission of the British Standards Institution. Complete copies of the documents can be obtained from the British Standards Institution, 389 Chiswick High Road, London W4 4AL.

Part 1
Health and Safety Management and Administration

Chapter 1
Principal Legal Requirements

The law on health and safety at work has, like much protective legislation, developed in a fragmented way over the last two centuries. In many cases, its growth has been brought about as a result of public outcry at, for instance, the appalling conditions under which children worked in the Lancashire textile mills following the Industrial Revolution. Indeed, the first statute, the Health and Morals of Apprentices Act 1802, was passed to combat this state of affairs in the textile industry. This Act limited the working hours of apprentices and established minimum standards of environmental control in terms of heating, ventilation and humidity control. Enforcement was by means of visitors to factories appointed by local magistrates.

The Factory Act 1833 resulted in the appointment of four factory inspectors with specific powers of entry to factories and means of enforcement. However, it was not until 1864 that such powers were extended beyond the textile industry to the matchmaking and pottery industries and, subsequently, with the Factory and Workshop Act 1878, to other forms of manufacturing industry. A final consolidating Act, the Factory and Workshop Act 1901, formed the basis for much of the current protective health and safety legislation. It enabled the Minister or Secretary of State to make specific Regulations covering a wide range of relatively dangerous industrial processes and practices.

Modern health and safety legislation commenced with the Factories Act 1937, the predecessor of the current Factories Act 1961. Similar legislation came into operation about this time dealing with mines and quarries, offices, shops and railway premises, and agriculture.

The Report of the Committee on Safety and Health at Work (Cmnd 5034) was published in July 1972 (The Robens Report). This Report proposed substantial changes in the law and administration of occupational health and safety and was instrumental in the passing of the Health and Safety at Work Act 1974, described in the following pages.

Since 1993 European Directives have had a significant effect on UK health and safety legislation.

Health and Safety at Work Act (HASAWA) 1974

1. Duties of employers
It is the duty of every employer, so far as is reasonably practicable, to

ensure the health, safety and welfare at work of all his employees (HASAWA Section 2(1)). In particular, this includes:

(a) the provision and maintenance of plant and systems of work that are safe and without risks to health (HASAWA Section 2(2) (a));
(b) arrangements for ensuring the safety and absence of health risks in connection with the use, handling, storage and transport of articles and substances (HASAWA Section 2(2) (b));
(c) the provision of such information, instruction, training and supervision as is necessary to ensure the health and safety at work of employees (HASAWA Section 2(2) (c));
(d) the maintenance of any place of work under the employer's control in a condition that is safe and without risks to health, and the provision and maintenance of means of access and egress from it that are safe and without risks to health (HASAWA Section 2(2) (d)); and
(e) the provision and maintenance of a working environment for his employees that is safe, without risks to health and adequate as regards facilities and arrangements for their welfare at work (HASAWA Section 2(2) (e)).

2. Duties of employees

It is the duty of every employee while at work:

(a) to take reasonable care for the health and safety of himself and of other persons who may be affected by his acts or omissions at work; and
(b) as regards any duty or requirement imposed on his employer or any other person by or under any of the relevant statutory provisions, to co-operate with him so far as is necessary to enable that duty or requirement to be performed or complied with (HASAWA Section 7).

No person shall intentionally or recklessly interfere with or misuse anything provided in the interests of health, safety or welfare in pursuance of any of the relevant statutory provisions (HASAWA Section 8).

3. Duties of employers to persons other than their employees

Every employer must conduct his undertaking in such a way as to ensure, so far as is reasonably practicable, that persons not in his employment, eg contractors, are not exposed to risks to health or safety (HASAWA Section 3(1)).

4. Duties of occupiers of premises to persons other than their employees

Section 4 of the HASAWA has effect for imposing on persons duties in relation to those who:

(a) are not their employees, but
(b) use non-domestic premises made available to them as a place of work or as a place where they may use plant or substances provided for their use there,

and applies to premises so made available and other non-domestic premises used in connection with them.

In this case the person or persons in control of the premises must take such measures to ensure, so far as is reasonably practicable, that the premises, all means of access thereto or egress therefrom, and any plant or substances in the premises or, as the case may be, provided for use there, is or are safe and without risks to health.

The protection under this section extends to visitors who are:

(a) workers, eg employees of a contract land clearance firm, and
(b) visitors to agricultural premises, eg members of the public invited to pick fruit.

5. Duties of designers, manufacturers, importers and suppliers

Section 6 of the HASAWA originally placed specific duties on the designers, manufacturers, importers and suppliers of articles and substances used at work. These duties were amended by Section 36 and Schedule 3 of the Consumer Protection Act (CPA) 1987 whereby, in the case of *articles* for use at work (see definition below), such persons must:

(a) ensure, so far as is reasonably practicable, that the article is so designed and constructed that it will be safe and without risks to health at all times when it is being set, cleaned, used or maintained by a person at work;
(b) carry out or arrange for the carrying out of such testing and examination as may be necessary for the performance of the duty imposed on him by the preceding paragraph;
(c) take such steps as are necessary to secure that persons supplied by that person with the article are provided with adequate information about the use for which the article is designed or has been tested and about any conditions necessary to ensure that it will be safe and without risks to health at all such times as are mentioned in paragraph (a) above and when it is being dismantled or disposed of; and

(d) take such steps as are necessary to ensure, so far as is reasonably practicable, that persons so supplied are provided with all such revisions of information provided by them by virtue of the preceding paragraph as are necessary by reason of its becoming known that anything gives rise to a serious risk to health or safety.

In the case of *substances* for use at work, such persons have a duty to:

(a) ensure, so far as is reasonably practicable, that the substance will be safe and without risks to health at all times when it is being used, handled, processed, stored or transported by a person at work or in premises to which Section 4 applies;
(b) carry out or arrange for the carrying out of such testing and examination as may be necessary for the performance of the duty imposed on him by the preceding paragraph;
(c) take such steps as are necessary to secure that persons supplied by that person with the substance are provided with adequate information about any risks to health or safety to which the inherent properties of the substance may give rise, about the results of any relevant tests which have been carried out on or in connection with the substance and about any conditions necessary to ensure that the substance will be safe and without risks to health at all such times as are mentioned in paragraph (a) above and when the substance is being disposed of; and
(d) take such steps as are necessary to secure, so far as is reasonably practicable, that persons so supplied are provided with all such revisions of information provided by them by virtue of the preceding paragraph as are necessary by reason of it becoming known that anything gives rise to a serious risk to health or safety.

Under the CPA, 'articles for use at work' and 'substances' are defined thus:

Article for use at work, means:

(a) any plant designed for use or operation (whether exclusively or not) by persons at work, or who erect or install any article of fairground equipment, and
(b) any article designed for use as a component in any such plant or equipment.

Substance, means any natural or artificial substance (including micro-organisms) intended for use (whether exclusively or not) by persons at work.

Approved Codes of Practice

The need to provide elaboration on the implementation of regulations is recognised in Section 16 of the HASAWA, which gives the Health and Safety Commission (HSC) power to prepare and approve codes of practice on matters contained not only in regulations but in Sections 2 to 7 of the Act. Before approving a code, the Health and Safety Executive (HSE), acting for the HSC, must consult with any interested body.

An Approved Code of Practice (ACOP) has a special legal status similar to the Highway Code. No one can be prosecuted for an infringement of an ACOP as such, but if someone is prosecuted for any breach of health and safety law to which an ACOP applies, the ACOP is admissible in evidence, and if the guidance it contains has not been followed, the defendant would need to be able to prove that he had fulfilled the legal requirements in some other way.

Thus an ACOP is a quasi-legal document and, although non-compliance does not constitute a breach, if the contravention of the Act or regulations is alleged, the fact that the ACOP was not followed would be accepted in court as evidence of failure to do all that was *reasonably practicable*. A defence would be to provide that *works of equivalent nature* had been carried out or something equally good or better had been done.

Examples of ACOPs are:

Control of asbestos at work (Control of Asbestos at Work Regulations 1987)
Control of substances hazardous to health (Control of Substances Hazardous to Health Regulations 1994)
Safety of pressure systems (Pressure Systems and Transportable Gas Containers Regulations 1989)

HSE Guidance Notes

The HSE issues Guidance Notes in some cases to supplement the information in both regulations and ACOPs. These Guidance Notes have no legal status and are purely of an advisory nature.

HSE Guidance Notes fall into six categories:

1. General safety
2. Chemical safety
3. Environmental hygiene
4. Medical series
5. Plant and machinery
6. Health and safety.

Examples of Guidance Notes are:

EH40   Occupational exposure limits
MS20   Pre-employment health screening
HS(G)37 Introduction to local exhaust ventilation
PM41   Application of photoelectric safety systems to machinery
GS20   Fire precautions in pressurised workings.

Powers of inspectors

An inspector appointed under the Act, eg Factories Inspector, Environmental Health Officer, Agricultural Health and Safety Inspector, has the following powers:

(a) to enter premises at any reasonable time, or at any time (day or night) if he has reason to believe a dangerous situation exists;

(b) where an inspector has reason to believe he may be obstructed from entering premises in the exercise of his duties, he may enter the premises accompanied by a police constable; he may also take with him any other person authorised by the enforcing authority, together with any equipment and materials he may need;

(c) to make those examinations and investigations he considers necessary to determine whether there has been a breach of the law;

(d) to direct that premises, or part of a premises, are left undisturbed for so long as is reasonably necessary;

(e) to take measurements and photographs and make such recordings as he considers necessary for the purpose of such investigations or examinations;

(f) to take samples of any articles or substances found in the premises, and of the atmosphere in or in the vicinity of any such premises;

(g) where he has reason to believe that any article or substance has caused or is likely to cause danger to health and safety, to arrange for it to be dismantled or subjected to any process or test;

(h) to take possession of and detain any article or substance for so long as is necessary, either for the purpose of examination or test, to ensure that it is not tampered with before his examination of it is completed, or to ensure that it is available for use as evidence in any subsequent proceedings;

(i) to question any person who may have information relevant to his investigations, either alone or in the presence of another person whom he has invited or allowed to be present, and require that person to answer questions and to sign a declaration of the truth of his answers;

(j) to require the production of, inspect and take copies of or any entry in, any books or documents which are required under the statutory provisions to be kept or which may be necessary for him to see for the purposes of any examination or investigation;

(k) to demand such facilities and assistance as may be necessary to enable him to exercise any of the above-mentioned powers; and

(l) to assume any other powers necessary to enable him to carry into effect the relevant statutory provisions.

If a person in charge of a premises requests to be present at the time, an inspector may not dismantle or subject to any process or test any article or substance other than in the presence of that person. Moreover, where an inspector takes possession of an article or substance, he must leave a notice with a responsible person giving particulars sufficient to identify it.

Where an inspector has reasonable cause to believe that any article or substance found in any premises which he has power to enter is a cause of imminent danger of serious personal injury, he may seize that article or substance and cause it to be rendered harmless. However, where practicable, the inspector must, first of all, take a sample of same and give to a responsible person at the premises a portion of the sample marked in a manner sufficient to identify it, together with a copy of his written report.

In situations where an inspector questions persons mentioned in (i) above, the answers given by that person in the course of the interrogation are not admissible in evidence in any subsequent proceedings against that person.

Improvement and prohibition notices

An inspector appointed under the HASAWA may serve two types of notice.

Improvement notices

Where an inspector is of the opinion that a person:

(a) is contravening one or more of the relevant statutory provisions, or

(b) has contravened one or more of these provisions in circumstances that make it likely that the contravention will continue or be repeated,

he may serve an improvement notice on that person requiring that he remedy the contravention(s) or, as the case may be, the matters occasioning it within such period (ending not earlier than the period within which an appeal may be brought) as may be specified in the notice.

Prohibition notices

Where an inspector is of the opinion that a work activity involves or will involve a risk of serious personal injury he may serve a prohibition notice on the owner and/or occupier of the premises or the person having control of that activity. Such a notice will direct that the specified activities in the notice shall not be carried on by or under the control of the person on whom the notice is served unless certain specified remedial measures have been complied with.

It should be noted that it is not necessary that an inspector believes that a legal provision is being or has been contravened. A prohibition notice is served where there is an immediate threat to life and in anticipation of danger.

A prohibition notice may have immediate effect after its issue by the inspector. Alternatively, it may be deferred, thereby allowing the person time to remedy the situation, carry out works etc. The duration of a deferred prohibition notice is stated on the notice.

Failure to comply with notices

In the case of both an improvement notice and a prohibition notice, where there is failure to comply within the time specified, or, in the event of an appeal against the notice, after the expiry of any extra time allowed for compliance by a tribunal, the person concerned can be prosecuted. Where a person is convicted of an offence specified in an improvement or prohibition notice and the contravention is continued after the conviction, he may be found guilty of a further offence and liable on summary conviction to a fine not exceeding £200 for each day on which the contravention is continued, in addition to the original maximum fine of £20,000 for each contravention specified.

The systems for enforcement under the HASAWA are summarised in figures 1, 2 and 3.

Appeals against notices

Where a person lodges an appeal against an improvement notice, the operation of that notice is automatically suspended. In the case of a prohibition notice, however, the requirements of the notice continue to apply, unless a tribunal has directed otherwise, in which case the suspension operates from the date directed by the tribunal.

Prosecution

Prosecution is frequently the outcome of failure to comply with an improvement or prohibition notice. Conversely, an inspector may simply institute legal proceedings without service of a notice. Cases are

WHAT POWERS HAVE HEALTH AND SAFETY INSPECTORS?

- To enter premises; make investigations; take samples/photos; ask questions, etc. (Section 20)
- to serve *improvement notices* on a person:
  - where there is breach of regulations,
  - with a minimum time limit of 21 days,
  - if employer appeals to industrial tribunal, notice void, pending appeal. (Section 21)
- To serve *prohibition notices* on persons controlling activities:
  - where there is risk of serious personal injury,
  - with immediate effect (or deferred where necessary),
  - if employer appeals, notice stays on pending appeal.
  (Section 22)
- To *seize and destroy articles/substances* that may cause serious personal injury. (Section 25)
- To give the following *information to safety representatives* under Section 28(8):
  - factual information on hazards, etc,
  - details of what the employer has been asked to do, eg notices.

WHAT ARE THE PENALTIES FOR BREAKING THE ACT?

- At magistrates courts:
  - maximum fine of £20,000.
- At jury trials (indictment)
  - unlimited fine, and/or
  - up to two years in prison for some offences listed in Section 33(4),
  - £200 per day for failing to comply with a notice (prohibition or improvement). (Section 33)

DOES THE ACT APPLY TO CROWN PROPERTY?

Yes – Part I, including general duties, 'binds the Crown'.
(Section 48.1)

CAN THE ACT BE ENFORCED AGAINST THE CROWN?

No – at least not through the courts, nor through enforcing improvement and prohibition notices. (Section 48.1)

CAN EMPLOYEES OF THE CROWN BE PROSECUTED?

Yes – Section 48(2), 36(2) and Section 37 allow prosecution against Crown employees where they are at fault.

Figure 1 Powers of health and safety inspectors

Notice	Circumstances		When notice takes effect	Effect of appeal (Section 24*)	Person on whom notice is served
	Contravention of a 'relevant statutory provision'	Risk involved			
Improvement (Section 21)	Must have been one and it is likely that it will be continued	No risk specified and covers cases where there is no risk	When specified but not earlier than 21 days after issue	Suspends the notice until appeal is determined or the appeal is withdrawn	Person contravening provision
Immediate prohibition notice (Section 22)	Not necessary	Where there is or will be immediate risk of serious personal injury **	Immediate	No suspension unless the industrial tribunal rules otherwise	Person under whose control the activity is carried on or by whom it is carried on
Deferred prohibition notice (Section 22)	Not necessary	Risk of serious personal injury not imminent	At the end of the period specified in the notice	As in 'immediate' notice	As in 'immediate' notice

* Appeals against notices are dealt with in regulations made under Section 24 as follows:
- The Industrial Tribunals (Improvement and Prohibition Notices Appeals) Regulations 1974 (SI 1974 No 1925) – for notices served in England and Wales, and
- The Industrial Tribunals (Improvement and Prohibition Notices Appeals) (Scotland) Regulations 1974 (SI 1974 No 1926) – for notices served in Scotland.

** 'Personal injury' includes any disease and any impairment of a person's physical or mental condition (Section 53).

Figure 2 Summary of features of notices under the Health and Safety at Work Act 1974

normally heard in a Magistrates Court, but there is also provision in the HASAWA on indictment. Much depends upon the gravity of the offence.

Corporate liability

Where an offence under any of the relevant statutory provisions (eg Regulations made under HASAWA 1974) committed by a body corporate is proved to have been committed with the consent or connivance of, or to have been attributable to any neglect on the part of, any director, manager, secretary or other similar officer of the body corporate or a person who was purporting to act in any such capacity, he as well as the body corporate shall be guilty of that offence and shall be liable to be proceeded against and punished accordingly. (Sec 37(1) HASAWA).

This means, fundamentally that:

(a) where an offence is committed through neglect or omission by a constituted board of directors, the company itself can be prosecuted as well as the directors individually who may have been to blame;

(b) where an individual functional director is guilty of an offence, he can be prosecuted as well as the company; and

(c) a company can be prosecuted even though the act or omission resulting in the offence was committed by a junior official or executive, or even a visitor to the company premises.

Other 'corporate' persons eg personnel managers, training officers, chief engineers, health and safety advisers, may also be guilty of offences. Section 36 of the HASAWA deals with such offences thus:

Where the commission by any person of an offence under any of the relevant statuory provisions is due to the act or default of some other person, that other person shall be guilty of the offence, and a person may be charged with and convicted of the offence . . . whether or not proceedings are taken against the first mentioned person.

Joint consultation with employees

The Safety Representatives and Safety Committees Regulations (SRSCR) 1977 provide for the appointment in prescribed cases by recognised trade unions of safety representatives from among the employees. Those selected represent the workforce in consultation with the employer with a view to promoting and developing measures to ensure the health and safety at work of employees, and in checking the effectiveness of such measures.

HEALTH AND SAFETY EXECUTIVE   Serial No. I

Health and Safety at Work etc. Act 1974, Sections 21, 23 and 24

IMPROVEMENT NOTICE

Name and address (See Section 46)

(a) Delete as necessary

(b) Inspector's full name

(c) Inspector's official designation

(d) Official address

(e) Location of premises or place and activity

(f) Other specified capacity

(g) Provisions contravened

To . . . .
. . . .
(a) Trading as . . . .
(b) . . . .
one of (c) . . . .
of (d) . . . .
. . . . Tel No . . . .
hereby give you notice That I am of the opinion that . . . .
(e) . . . .
you, as (a) an employer/s self employed person/s person wholly or partly in control of the premises
(f) . . . .
   (a) are contravening/have contravened in circumstances that make it likely that the contravention will continue or be repeated
. . . .
. . . .
(g) . . . .
. . . .
The reasons for my said opinion are:- . . . .
. . . .
. . . .
and I hereby require you to remedy the said contraventions or, as the case may be, the matters occasioning them by

(h) Date

(h) . . . .
(a) In the manner stated in the attached schedule which forms part of the notice.
Signature . . . . Date . . . .
Being an inspector appointed by an instrument in writing made pursuant to Section 19 of the said Act and entitled to issue this notice.
(a) An Improvement notice is also being served on
. . . .
of . . . .

LP1

related to the matters contained in this notice.

Figure 3 Improvement and prohibition notices

HEALTH AND SAFETY EXECUTIVE   Serial No. I

Health and Safety at Work etc. Act 1974, Sections 22–24 Serial No. P

PROHIBITION NOTICE

Name and
address (See
Section 46)

(a) Delete as
    necessary

(b) Inspector's
    full name

(c) Inspector's
    official
    designation

(d) Official
    address

To . . . .

. . . .

(a) Trading as . . . .

(b) . . . .

one of (c) . . . .

of (d) . . . .

. . . . tel no . . . .

hereby give you notice that I am of the opinion that the following
activities

namely:- . . . .

. . . .

. . . .

which are (a) being carried on by you/about to be carried on
by you/under your control

(e) Location
    of activity

at (e) . . . .

involve, or will involve (a) a risk/an imminent risk, of serious
personal injury. I am further of the opinion that the said matters
involve contraventions of the following statutory provision:- . . . .

. . . .

. . . .

. . . .

because . . . .

. . . .

. . . .

and I hereby direct that the said activities shall not be carried
on by you or under your control (a) immediately/after

(f) Date

(f) . . . .

unless the said contraventions and matters included in the
schedule, which forms part of this notice, have been remedied.

Signature . . . . Date . . . .

being an inspector appointed by an instrument in writing made
pursuant to Section 19 of the said Act and entitled to issue this
notice.

LP2

1. Safety representatives

A safety representative is a person appointed by his trade union to represent workers in consultations with the employer on all matters relating to health and safety at work. He has the following specific functions, following notification to the employer in writing of his appointment:

(a) to investigate potential hazards and causes of accidents at the workplace;

(b) to investigate complaints from employees concerning health and safety risks at work;

(c) to make representation to the employer on matters arising out of (a) and (b) above and on general matters affecting the health, safety and welfare of employees;

(d) to carry out certain inspections:
  - Of the workplace, after giving reasonable notice to the employer
  - of a relevant area following a reportable accident or scheduled dangerous occurrence, or where a reportable disease is contracted, if it is safe to do so and in the interests of the employees represented
  - of documents relating to the workplace or employees which the employer is required to maintain;

(e) to represent his group of employees in consultation with inspectors appointed under the Act, and to receive information from them;

(f) to attend meetings of safety committees.

2. Safety committees

Where requested in writing by at least two safety representatives the employer must form a safety committee (SRSCR). When establishing a safety committee, the employer must:

(a) consult with both the safety representatives making this request, and with representatives of the trade union whose members work in any workplace where it is proposed that the committee will function;

(b) post a notice stating the composition of the committee and the workplace(s) to be covered by it, in a place where it can easily be read by employees;

(c) establish the committee within three months following the request for its formation.

It is the employer's prerogative to establish the objectives, role and function of the safety committee, together with aspects such as frequency of meeting, rotation of Chairman, minutes, agenda, etc.

It should be appreciated that safety committees have a significant role in reducing accidents and occupational ill health. They are a forum for discussion on a wide range of topics, an important system for dissemination of information and can be extremely effective in maintaining good standards of communication between employers and employees.

3. Non-unionised employees
Under the Health and Safety (Consultation with Employees) Regulations 1996 employers must consult any employees who are not covered by the Safety Representatives and Safety Committees Regulations. This may be by direct consultation with employees or through representatives elected by the employees they are to represent.

HSE Guidance
HSE Guidance accompanying the Regulations details:

(a) which employees must be involved;
(b) the information they must be provided with;
(c) procedures for the election of representatives of employee safety;
(d) the training, time off and facilities they must be provided with; and
(e) their functions in office.

Statements of Health and Safety Policy

Section 2(3) of the HASAWA imposes a duty on every employer of five or more persons to prepare, and bring to the notice of his employees, a written statement of his general policy with respect to the health and safety at work of his employees. Guidance is provided in Leaflet No HSC 6 *Guidance notes on employers' policy statements for Health and Safety at Work* and *Writing your Health and Safety Policy Statement – how to prepare a safety policy statement for a small business*, both available from HMSO.

Content of the Policy Statement
Essentially, a Statement of Health and Safety Policy should be in written form and signed and dated by the owner, occupier or person having control of a business. It should consist of three parts principally, and be subject to regular revision according to individual circumstances, eg change of individual responsibilities. It is standard practice, further, to incorporate a number of appendices dealing with specific aspects of the Policy Statement, in particular the individual responsibilities of all levels of management and workers for health and safety. The three principal parts of a Statement of Health and Safety Policy are as follows.

1. A general statement of intent

This should outline in broad terms the company's overall philosophy in relation to the management of health and safety. It should also include broad reference to the responsibilities of directors, line management and employees.

2. Organisation

This part is concerned with people and their duties, and outlines the chain of command in terms of health and safety management, in particular individual accountabilities, the system for monitoring implementation of the Policy, the role and function of trade union safety representatives and of the safety committee, and a management chart showing the lines of responsibility from managing director or chief executive downwards.

3. Arrangements

This part deals with the systems and procedures for ensuring appropriate standards of safety, health and welfare, including the practical arrangements for their implementation.

Aspects to be incorporated in the 'Arrangements' include, for instance, the system for health and safety training, control of the working environment, procedures for operating safe systems of work, machine guarding, housekeeping procedures, noise control, dust control, fire protection procedures, health surveillance arrangements, systems for reporting, recording and investigation of accidents, ill health and dangerous occurrences, emergency procedures, eg in the event of fire, and safety monitoring systems. This part of the Policy Statement, above all, must be a 'living document' which is subject to regular revision and updating.

Fundamentally, a Statement of Health and Safety Policy is specific to a particular premises or business operation. It must be carefully thought out, there must be consultation with the workforce during its preparation if it is to be successful, and, above all, everyone must be fully aware of their responsibilities towards themselves and others with the principal aim of preventing accidents, ill health and other loss-producing incidents.

Appendices to the Policy Statement

Where specific attention to a particular aspect is necessary, it is common practice to add a number of appendices to the Policy Statement. Such appendices could include:

(a) general and specific legislation applying to the business;
(b) individual responsibilities of management and employees;

(c) specific company policies relating to, for instance, Aids, smoking at work, pre-employment health examinations and other forms of health surveillance, the provision and use of certain forms of personal protective equipment, eg eye protection, first aid procedures;

(d) the hazards that could be encountered and the precautions necessary on the part of workers, eg from machinery, dangerous substances, or the use of ladders;

(e) joint consultation procedures;

(f) the system for providing health and safety information to employees;

(g) fatal and major injury accident procedure;

(h) procedures to protect visitors, eg contractors, and members of the public from risks arising from work activities.

The hierarchy of duties

The duties of people at work under the FA, HASAWA, etc, vary. Certain legal duties on employers and others may be of an absolute nature, or qualified by the terms 'where practicable' or 'so far as is reasonably practicable'.

'Absolute' duties: Where there is a high degree of risk of death or serious injury if safety precautions are not taken, as in the case of certain classes of machinery, the duty may well be of an absolute nature. Such duties are laid down in the FA in the case of, for instance, prime movers. Sec 12(1) requires that 'every flywheel directly connected to any prime mover and any moving part of any prime mover . . . shall be securely fenced whether the flywheel or prime mover is situated in an engine house or not'.

'Practicable' means more than physically possible. Thus in Adsett *v* Steel Founders Ltd, the Judge said that the measures must be able to be carried out 'in the light of current knowledge and invention'. Moreover, it was pointed out that 'practicable' implies a higher standard of care than the term 'reasonably practicable'.

'Reasonably practicable' means a comparison must be made between, on the one hand, the extent of the risk and, on the other, the sacrifice (costs, time and effort) in taking the measures necessary to avert the risk. If it can be shown that there is a disparity between the degree of risk and sacrifice involved, then the defendant has discharged the burden of proof.

Both the above expressions are concerned with the burden of proof, the former imposing a higher standard than the latter.

'All reasonable precautions and all due diligence'

An important new defence, taken from food safety legislation, has been incorporated in recent health and safety legislation, such as the Control of Substances Hazardous to Health (COSHH) Regulations 1994, Electricity at Work Regulations 1989 and the Pressure Systems and Transportable Gas Containers Regulations 1989. This defence makes provision for a person charged with an offence to plead that he took '**all** reasonable precautions or steps and exercised **all** due diligence' to prevent the commission of the offence. The significance of the word 'all' in each case should be noted, and if such a defence is to be successful, proof of management and operation of many of the following systems and procedures, and supported by appropriate documentation eg internal codes of practice, is necessary.

1. A well-written Statement of Health and Safety Policy, which clearly identifies individual responsibilities of all concerned.
2. Preventive maintenance schedules.
3. Cleaning schedules.
4. Procedures for providing information, instruction and training to staff, contractors and visitors, in particular, induction and orientation training procedures.
5. Formally documented safe systems of work, together with Permit to Work systems.
6. Identification of persons classed as 'competent persons' together with their specific responsibilities.
7. Safety monitoring procedures eg safety audits.
8. Accident and incident reporting, recording and investigation procedures.
9. Occupational health and hygiene procedures.
10. Environmental control systems eg noise control.
11. Procedures for vetting the relative safety of new machinery, plant, equipment and potentially dangerous substances.
12. Procedures for regulating the activities of contractors.
13. Fire protection and evacuation procedures.
14. Procedures for liaison with enforcement agencies eg Health and Safety Executive, and action to be taken following the service of an Improvement or Prohibition Notice.
15. Health and safety promotional activities, eg Health and Safety Awards.

Documentation of the above is extremely important, and many of the above policies, systems and procedures should be incorporated in a company Health and Safety Manual which should be related to and referred to in the Statement of Health and Safety Policy.

Management of Health and Safety at Work Regulations (MHSWR) 1992
These Regulations, which are accompanied by an ACOP, implement the European Framework Directive 'on the introduction of measures to

encourage improvements in the safety and health of workers at work'. As such, they brought in a totally new management-orientated approach to health and safety law, including the duty on employers to take into account 'human capability' as regards health and safety when entrusting tasks to their employees. What is of particular significance is the absolute nature (ie 'shall') of the duties on employers compared with those under the HASAWA which are qualified by the phrase 'so far as is reasonably practicable'.

Under the MHSWR, every employer *shall* make a suitable and sufficient assessment of the risks to the health and safety of:

(i) his employees, noting risks to which they are exposed whilst at work; and

(ii) persons not in his employment, noting risks arising out of or in connection with his own conduct.

This is necessary to ensure compliance with the requirements and prohibitions imposed upon him by or under the relevant statutory provisions. Such risk assessments shall be reviewed where no longer valid or where there has been a significant change in related matters. Where more than five people are employed, the significant findings of the assessment and any group of employees identified by it as being specifically at risk, shall be recorded (in writing). (Regulation 3)

Regulation 4 fundamentally places an absolute duty on employers to manage their health and safety activities, and is quite simply stated thus:

'Every employer shall make and give effect to such arrangements as are appropriate, having regard to the nature of his activities and the size of his undertaking, for the effective *planning, organisation, control, monitoring and review* of the preventive and protective measures.' Again, where five or more people are employed, these arrangements must be recorded.

Where the risk assessment undertaken under Regulation 3 identifies the need for health surveillance, every employer shall ensure his employees are provided with same. (Regulation 5)

A very important feature of these Regulations is the duty of employers to appoint one or more 'competent persons' (see also p. 44) to assist him in undertaking the measures necessary to comply with the requirements and prohibitions imposed upon him by or under the relevant statutory provisions. (Regulation 6) Competent persons may be appointed from within the organisation or from outside, eg consultants. The employer shall ensure that the number of persons appointed as competent persons, the time available for them to fulfil their functions, and the means at their disposal are adequate, having regard to the size of the undertaking, the risks to which his employees are exposed and the distribution of those risks. External competent persons must also be adequately informed of any factors affecting health and safety. The employer must assess the competence of his competent person on the basis of his having 'sufficient training and experience or knowledge and other qualities to enable him properly to assist in undertaking the measures he needs to take to comply with the requirements and prohibitions imposed upon him by or under the relevant statutory provisions'.

Regulation 7 requires employers to establish written 'procedures for serious and imminent danger and for danger areas'. Fundamentally, an employer must predict the worst possible event that could arise in his undertaking, such as a major escalating fire, and plan emergency procedures around same. He must also appoint 'competent persons' for the purpose of implementing the emergency procedures. The risk assessment undertaken in accordance with Regulation 3 should give an indication of potential emergency situations and danger areas. The ACOP defines a 'danger area' as 'a work environment which must be entered by an employee where the level of risk is unacceptable without special precautions being taken'.

Under Regulation 8 every employer shall provide his employees with *comprehensible and relevant* information on the risks to their health and safety identified by the assessment, the preventive and protective measures, the emergency procedures, the competent persons nominated to implement the emergency procedures and the risks associated with shared workplaces.

In situations of shared workplaces, eg construction sites or where contractors may be operating in an existing workplace, every such employer must co-operate with other employers and assist in the co-ordination of safety measures, together with informing other employers concerned of any risks arising out of the work. (Regulation 9) In some cases, this may require the appointment of a health and safety co-ordinator, particularly where major construction activities are involved. Regulation 10 further requires employers to provide comprehensible information to the employers of external employees, eg cleaning contractors, on the risks arising and the measures taken by the first employer to prevent same arising.

Regulation 11 places an absolute duty on employers to take into account the human capabilities of employees as regards health and safety when entrusting them with tasks, and to ensure that employees are provided with adequate health and safety training:

(a) on recruitment;
(b) on being exposed to new or increased risks because of
    (i) their being transferred or given a change in responsibilities;
    (ii) the introduction of new work equipment or a change in use of existing work equipment;

(iii) the introduction of new technology; or
(iv) the introduction of a new system of work or a change respecting an existing system.

The above training must be repeated periodically where appropriate, must be adapted to take account of any new or changed risks, and must take place during working hours.

Under Regulation 12(1) the duties of employees under Section 7 of the HASAWA are clarified thus:

'Every employee shall use any machinery, equipment, dangerous substance, transport equipment, means of production or safety device provided to him by his employer in accordance both with any training in the use of the equipment concerned which has been received by him and the instructions respecting that use which have been provided to him by the said employer in compliance with the requirements and prohibitions imposed upon that employer by or under the relevant statutory provisions.'

Regulation 12(2) further places an absolute duty on employees to inform their employer or health and safety officer/adviser of any situation which represents a serious and immediate danger to health and safety and any matter which could be considered a shortcoming in the employer's protection arrangements for health and safety.

The situation relating to the question of temporary workers is covered in Regulation 13. Under this regulation an employer shall provide:

(a) any person he has employed under a fixed-term contract of employment; or
(b) any person employed in an employment business who is carrying out work in his undertaking;

with comprehensible information on:

(a) any special occupational qualifications or skills required to be held by that employee if he is to carry out his work safely; and
(b) any health surveillance required to be provided to that employee by or under the relevant statutory provisions.

Management of Health and Safety at Work (Amendment) Regulations 1994

These Regulations implemented the provisions of the European Directive on Pregnant Workers.

Duties of employers
The Regulations require employers to:

(a) assess the risks to the health and safety of women who are pregnant, have recently given birth, or who are breastfeeding;
(b) ensure that workers are not exposed to risks identified by the risk assessment, which would present a danger to their health or safety.

If, after taking whatever preventive action is reasonable, there is still a significant risk, which goes beyond the level of risk to be expected outside the workplace, then employers must take the following steps to remove the employee from that risk:

(a) temporarily adjust working conditions or hours of work;
(b) if that is not reasonable or would not avoid the risk, they should offer her alternative work if any is available;
(c) if that is not possible, they must give her paid leave from work for as long as is necessary to protect her health and safety or that of her child.

If a new or expectant mother works at night and has a medical certificate stating that night work could damage her health or safety, the employer must either offer her daytime work if any is available or, if that is not reasonable, give her paid leave for as long as necessary to protect her health or safety.

Health and Safety (Young Persons) Regulations 1997
These Regulations implement Articles 6 and 7 (the health and safety provisions) of the European Directive on the Protection of Young People at Work.
A young person is a person under the age of 18 years.
These regulations require the employer to:

(a) take particular account of young workers' lack of experience, absence of awareness of existing or potential risks or their immaturity, when they assess the risks to their health and safety; the assessment must now be made before the young person begins work and must address specific factors;
(b) take account of the risk assessment in determining whether the young person is prohibited from doing certain work; this requirement does not apply to young people over school leaving age where the work is necessary for their training and they are properly supervised and the risks are reduced to the lowest practicable level;

(c) inform parents or those with parental responsibility for school age children of the outcome of the risk assessment and the control measures introduced.

The regulations do not apply to young people doing occasional or short-term work in a family undertaking when that work is not harmful, damaging or dangerous. However, other health and safety legislation, such as the need to carry out risk assessments and take measures to control any risks identified, applies in these circumstances.

Work away from base
A substantial number of people work away from their normal base of operations, such as building contractors, people involved in the installation and servicing of equipment, drivers and company representatives. As such, they are exposed to a wide range of hazards though, in most cases, their unfamiliarity with premises, processes and working practices. In some organisations around 20 to 25 per cent of accidents to staff take place on other people's premises.

The legal requirements relating to work on other people's premises are dealt with in the Occupiers' Liability Act 1957, Health and Safety at Work etc Act 1974 and the Management of Health and Safety at Work Regulations 1992.

Occupiers' Liability Act 1975
An occupier of premises owes a *common duty of care* to all *lawful* visitors in respect of dangers due to the *state* of the premises or *things done or omitted to be done* on them.

Health and Safety at Work etc Act 1974
An employer must conduct his undertaking in such a way as to ensure, so far as reasonably practicable, that *persons not in his employment* who may be affected thereby are not thereby exposes to risks to their health or safety. (Section 3)

Every person who has, to any extent, control of premises must ensure, so far as is reasonably practicable, that the premises, all means of access thereto and egress therefrom, and any plant or substances in the premises or provided for use there, is or are safe and without risks to health. (Section 4)

Management of Health and Safety at Work Regulations 1992
Regulation 10 deals with *persons working in host employers' undertakings*. Host employers must provide such persons with *appropriate instructions and comprehensible information* on the risks arising out of or in connection with his undertaking and, secondly, the measures taken in compliance with the requirements and prohibitions imposed upon him by or under

the relevant statutory provisions insofar as the said requirements and prohibitions relate to those employees.

Summary

The duties of employers and occupiers of premises towards non-employees working in their premises or undertaking can be summarised as follows:

(a) a general duty of care to all lawful visitors;
(b) a general duty not to expose non-employees to risks to their health or safety;
(c) a general duty on controllers of premises to provide safe premises, safe access and egress, safe plant and substances;
(d) a specific duty to provide instructions and information on risks and precautionary measures necessary by the non-employees or self-employed persons concerned.

Practical procedures to implement these requirements

1. Provision of written instructions to non-employees with regard to safe working practices generally.
2. Provision of comprehensible written information to non-employees on the hazards and precautions necessary.
3. Formal health and safety training sessions for all non-employees prior to commencing work on the host employer's premises.
4. Specification of the health and safety competence necessary for non-employees at the tender stage of contracts.
5. Operation of formal hazard reporting systems by non-employees.
6. Disciplinary procedures against non-employees for a failure to comply with written instructions and safety signs, including dismissal from the site or premises in serious cases of non-compliance.
7. General supervision of non-employees and regular meetings to reinforce the safety requirements for such persons.
8. Liaison with the external employers, and certification where necessary, to ensure the employees concerned have received the appropriate information, instruction and training necessary prior to commencing work in the host employer's undertaking.
9. Pre-tender and on-going site inspections by the external employer to ensure his employees are not exposed to risks to their health or safety.

Current trends in health and safety legislation

1. All modern health and safety legislation is largely driven by European Directives. For instance,
   (a) the Directive 'on the health and safety of workers at work' was

implemented in the UK as the Management of Health and Safety at Work Regulations 1992; and

(b) the Temporary and Mobile Construction Sites Directive was implemented in the UK as the Construction (Design and Management) Regulations 1994.

2. Regulations produced since 1992 do not, in most cases, stand on their own. They must be read in conjunction with the general duties imposed on employers under the Management of Health and Safety at Work Regulations 1992, in particular the duties relating to:

(a) risk assessment;
(b) the operation and maintenance of safety management systems;
(c) the appointment of competent persons;
(d) establishment and implementation of emergency procedures;
(e) provision of information which is comprehensible and relevant;
(f) co-operation, communication and co-ordination between employers in shared workplaces, eg construction sites, office blocks, industrial estates;
(g) provision of comprehensible health and safety information to employees from an outside undertaking;
(h) assessment of human capability prior to allocating tasks; and
(i) provision of health and safety training and instruction.

The employer's duties at common law

Common law (case law) is a special body of law that has developed over the centuries. It is based on the judgements of the courts, each judgement containing a judge's enunciation of the facts, a statement of the law applying to the case and his *ratio decidendi* or legal reasoning for the conclusion or finding that he has arrived at. These judgements are recorded in the various Law Reports and form a body of decisions or precedents which other courts must follow.

At common law, employers owe their employees a general duty to take reasonable care in order to avoid injuries, diseases and deaths occurring at work. In particular, employers must:

(a) provide a safe place of work with safe means of access and egress;
(b) provide and maintain safe appliances and equipment and plant for doing the work;
(c) provide and maintain a safe system of work; and
(d) provide competent people to undertake the work.

(Wilsons & Clyde Coal Co Ltd *v* English (1938) 2 AER 628)

All these common law duties were incorporated in Section 2 of the HASAWA under the general duties of the employer.

Negligence

The above common law duties feature in the general law of negligence and constitute specific aspects of the duty to take reasonable care. Thus, in order to prove negligence, a plaintiff must satisfy the following criteria, which are incorporated in the definition of 'negligence', ie:

(a) the existence of a duty of care owed by the defendant to the plaintiff, eg by the employer to the employee;
(b) breach of that duty; and
(c) injury, damage or loss resulting from or caused by that breach.

(Lochgelly Iron & Coal Co Ltd *v* M'Mullan (1934) AC 1)

Courts and Tribunals

There are two distinct systems whereby the courts deal with criminal and civil actions respectively. However, some courts have both criminal and civil jurisdiction. The following list outlines the court system operating in the UK.

1.   Magistrates Court

This is the lowest of the courts in England and Wales and deals mainly with criminal matters. Its jurisdiction is limited. Magistrates determine and sentence for many of the less serious offences. They also hold preliminary examinations into other offences to ascertain whether the prosecution can show a prima facie case on which the accused may be committed for trial at a higher court. The Sheriff Court performs a parallel function in Scotland, although procedures differ from those of the Magistrates Court.

2.   Crown Court

Serious criminal charges and cases where the accused has the right to jury trial are heard on indictment in the Crown Court before a judge and jury. This court also hears appeals from magistrates courts.

3.   County Courts

These courts operate on an area basis and deal in the first instance with a wide range of civil matters. They are limited, however, in the remedies that can be applied. Cases are generally heard by circuit judges or registrars, the latter having limited jurisdiction.

4.   High Court of Justice

More important civil matters, because of the sums involved or legal complexity, will start in the High Court of Justice before a High Court judge. The High Court has three divisions:

(a) Queen's Bench – deals with contract and torts.
(b) Chancery – deals with matters relating to areas such as; land, wills, bankruptcy, partnerships and companies.
(c) Family – deals with matters involving issues such as; adoption of children, marital property and disputes.

In addition, the Queen's Bench Division hears appeals on matters of law:

(a) from the magistrates courts and from the Crown Court on a procedure called 'case stated'; and
(b) from some tribunals, for example the finding of an industrial tribunal on an enforcement notice under the Health and Safety at Work Act.

It also has some supervisory powers over the lower courts and tribunals. These can be exercised if the latter exceed their jurisdiction, fail to undertake their responsibilities properly or neglect to carry out any of their duties.

The High Court, the Crown Court and the Court of Appeal are known as the Supreme Court of Judicature.

5. The Court of Appeal
The Court of Appeal has two divisions:

(a) the Civil Division, which hears appeals from the county courts and the High Court; and
(b) the Criminal Division, which hears appeals from the Crown Court.

6. The House of Lords
The Law Lords deal with important matters of law only. This can only follow an appeal to the House of Lords from the Court of Appeal and in restricted circumstances from the High Court.

7. European Court of Justice
This is the supreme law court, whose decisions on interpretation of European Union law are sacrosanct. Such decisions are enforce- able through the network of courts and tribunals in all member states.

Cases can only be brought before this court by organisations, or by individuals representing organisations.

Tribunals
Industrial tribunals were first established under the Industrial Training Act of 1964 to deal with appeals against industrial training levies by employers. They now cover many industrial matters. These include,

industrial relations issues, cases involving unfair dismissal, equal pay and sex discrimination.

Composition

Each tribunal consists of a legally qualified chairman appointed by the Lord Chancellor and two lay members, one from management and one from a trade union. These representatives are selected from panels maintained by the Department of Employment following nominations from employers' organisations and trade unions.

Decisions

When all three members of a tribunal are sitting the majority view prevails.

Complaints relating to health and safety matters

Industrial tribunals deal with the following employment/health and safety issues:

(a) appeals against improvement and prohibition notices served by enforcement officers;

(b) time off for the training of safety representatives (Safety Representa- tives and Safety Committees Regulations 1977, Reg 11(1)(a));

(c) failure of an employer to pay a safety representative for time off for undertaking his functions and training (Safety Representatives and Safety Committees Regulations 1977, Reg 11(1)(b));

(d) failure of an employer to make a medical suspension payment (Employment Protection (Consolidation) Act 1978, Sec 22); and

(e) dismissal, actual or constructive, following a breach of health and safety law, regulation and/or term of employment contract.

The Health and Safety Information for Employees Regulations 1989

These Regulations require information relating to health, safety and welfare to be furnished to employees by means of posters or leaflets in the form approved and published for the purposes of the Regulations by the HSE (Regulations 3 and 4). Copies of the form of poster or leaflets approved in this way may be obtained from HMSO.

The Regulations also require the name and address of the enforcing authority and the address of the employment medical advisory service to be written in the appropriate space on the poster (Regulation 5 (1)); and where the leaflet is given the same information should be specified in a written notice accompanying it (Regulation 5 (3)).

This leaflet is reproduced in Figure 4 (see page 31).

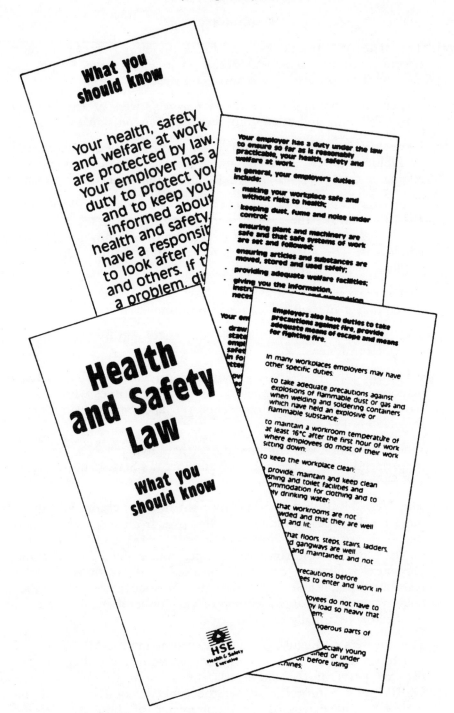

Figure 4   Health and safety information leaflet

Summary

1. The HASAWA places specific duties on employers, employees, occupiers of premises, and designers, manufacturers, importers and suppliers of articles and substances used at work.

2. Inspectors appointed under the Act have powers of entry at any reasonable time and at any time in special circumstances.

3. Inspectors have powers to serve both improvement notices and prohibition notices, to take samples, to require dismantling and testing, and to seize unsafe articles and/or substances.

4. The Safety Representatives and Safety Committees Regulations 1977 provide for the appointment in prescribed cases by recognised trade unions of safety representatives from among the employees. The specific functions of safety representatives are outlined in the Regulations.

5. Where requested in writing by at least two safety representatives, the employer must form a safety committee.

6. The employer establishes the role, objectives and functions of the safety committee.

7. Under the Health and Safety (Consultation with Employees) Regulations 1996 employers must consult any employees who are not covered by the Safety Representatives and Safety Committees Regulations 1977.

8. Where an employer employs more than five persons, he must prepare and bring to the notice of employees a Statement of Health and Safety Policy.

9. The MHSWR 1992 bring in important new provisions to implement the European Framework Directive. It must be appreciated that all these new duties imposed on employers are of an 'absolute' or strict nature compared with duties under the HASAWA and previous Regulations, which are qualified by the term 'so far as is reasonably practicable'. The room for manoeuvre on the part of employers, when charged with an offence under these Regulations, has been substantially reduced.

10. The five sets of Regulations which accompany the MHSWR, and which should be read in conjunction with these Regulations, revoked parts of the old legislation, such as the FA and the OSRPA, and reinforce many of the general requirements of the MHSWR. The details of the requirements of these five sets of Regulations are covered in specific chapters.

11. Specific provisions are made for pregnant workers and young persons.

12. Where employees are working away from base, both the employer and host employer have specific responsibilities.

References

Fife, Judge Ian, and Machin, E A (1988) *Redgrave's Health and Safety in Factories* Butterworths, London

Health and Safety Commission (1976) *Guidance notes on employers' policy statements for Health and Safety at Work* HMSO, London

Health and Safety Commission (1977) *Safety Representatives and Safety Committees Regulations 1977* HMSO, London

Health and Safety Commission (1977) *Writing your Health and Safety Policy Statement – How to prepare a safety policy statement for a small business* HMSO, London

Health and Safety Commission (1992) *Management of Health and Safety at Work Regulations 1992 and Approved Code of Practice* HMSO, London

Health and Safety Executive (1980) *Effective Polices for Health and Safety* HMSO, London

Health and Safety Executive (1989) *Health and Safety Law: What you should know* HSE Information Centre, Sheffield

Health and Safety Executive (1991) *Successful Health and Safety Management (HS(G)65)* HMSO, London

Secretary of State for Employment (1972) *Report of the Committee on Safety and Health at Work* (Robens Report) (Cmnd 5034) HMSO, London

Stranks, J (1996) *Health and Safety Law* Pitman, London

Chapter 2
Health and Safety Management

What is management?

One definition is:

'The effective use of resources in the pursuit of organisational goals.'

'Effective' implies achieving a balance between the risk of being in business and the cost of eliminating or reducing those risks.

Management entails leadership, authority and co-ordination of resources, together with:

(a) planning and organisation;
(b) co-ordination and control;
(c) communication;
(d) selection and placement of subordinates;
(e) training and development of subordinates;
(f) accountability; and
(g) responsibility.

Management resources include:

(a) people – employees, managers;
(b) land and buildings; and
(c) capital; together with
(d) time; and
(e) management skills in co-ordinating the use of resources.

Health and safety management

Health and safety management is no different from other forms of management.

It covers:

(a) the management of the health and safety operation at national and local level – planning, organising, controlling, objective setting, establishing accountability and the setting of policy;
(b) measurement of health and safety performance on the part of individuals and specific locations; and

(c) motivating managers to improve standards of health and safety performance in those areas under their control.

Decision-making is a very important feature of the management process. This can be summarised thus:

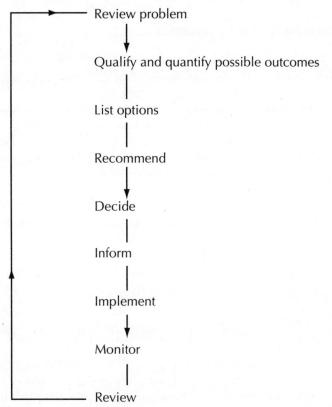

Review problem

Qualify and quantify possible outcomes

List options

Recommend

Decide

Inform

Implement

Monitor

Review

Management performance

Management is concerned with people at all levels of the organisation and human behaviour, in particular human personal factors such as attitude, perception, motivation, personality, learning and training. Communication brings these various behavioural factors together.

The management of health and safety is also concerned with organisational structures, the climate for change within an organisation, individual roles within the organisation and the problem of stress which takes many forms. Several questions must be asked at this stage.

1. Are managers good at managing health and safety?
2. What is their attitude to health and safety? Is it of a pro-active or reactive nature?

3. How do their attitudes affect the development of health and safety systems?
4. Is health and safety a management tool, or a problem thrust on to them by the enforcement agencies?

Health and safety management in practice
Whilst health and safety management covers many areas, there are a number of aspects which are significant, namely:

(a) the company Statement of Health and Safety Policy;
(b) procedures for health and safety monitoring and performance measurement;
(c) clear identification of the objectives and standards which must be measurable and achievable by the persons concerned;
(d) the system for improving knowledge, attitudes and motivation and for increasing individual awareness of health and safety issues, responsibilities and accountabilities;
(e) procedures for eliminating potential hazards from plant, machinery, substances and working practices through risk assessment, the design and operation of safe systems of work and other forms of hazard control; and
(f) measures taken by management to ensure legal compliance.

Information sources for health and safety
Formal (primary) sources

1. EU Directives
These are the Community instrument of legislation. Directives are legally binding on the governments of all member states who must introduce national legislation, or use administration procedures where applicable, to implement its requirements.

2. Acts of Parliament (Statutes)
Acts of Parliament can be innovatory, ie introducing new legislation, or consolidating, ie reinforcing, with modifications, existing law. Statutes empower the Minister or Secretary of State to make Regulations (delegated or subordinate legislation). Typical examples are the HASAWA and the Factories Act 1961.

3. Regulations (Statutory Instruments)
Regulations are more detailed than the parent Act, which lays down the framework and objectives of the system. Specific details are incorporated in Regulations made under the Act, eg the Control of Substances Hazardous to Health (COSHH) Regulations 1994 which were passed

pursuant to the HASWA. Regulations are made by the appropriate Minister or Secretary of State whose powers to do so are identified in the parent Act.

4. *Statutory Rules and Orders (SR & Os)*

These are the earlier equivalent of Statutory Instruments. They ceased to be published in 1948 but there are many applicable to occupational health and safety which are still in force, eg Chains, Ropes and Lifting Tackle (Register) Order 1938.

5. *Approved Codes of Practice*

The HSC is empowered to approve and issue Codes of Practice for the purpose of providing guidance on health and safety duties and other matters laid down in Statute or Regulations. A Code of Practice can be drawn up by the Commission or the HSE. In every case, however, the relevant government department, or other body, must be consulted beforehand and approval of the Secretary of State must be obtained. Any Code of Practice approved in this way is an Approved Code of Practice (ACOP).

An ACOP enjoys a special status under the HASWA. Although failure to comply with any provision of the code is not in itself an offence, that failure may be taken by a court, in criminal proceedings, as proof that a person has contravened the legal requirement to which the provision relates. In such a case it will be open to that person to satisfy a court that he or she has complied with the requirement in some other way.

Examples of ACOPs are those issued with the 'Control of Substances Hazardous to Health' (COSHH) Regulations entitled 'Control of Substances Hazardous to Health', 'Control of Carcinogenic Substances' and 'Control of Substances Hazardous to Health in Fumigation Operations'.

6. *Case law (common law)*

Case law is an important source of information. It is derived from common law because, traditionally, judges have formulated rules and principles of law as the cases occur for decision before the courts.

What is important in a case is the *ratio decidendi* (the reason for the decision). This is binding on courts of equal rank who may be deciding the same point of law. Ratio decidendi is the application of such an established principle to the facts of a given case, for instance, negligence consists of omitting to do what a 'reasonable man' would do in order to avoid causing injury to others.

Case law is found in law reports, for example, the All England Law Reports, the Industrial Cases Reports, the current *Law Year Book* and in professional journals, eg *Law Society Gazette, Solicitors Journal*. In addition, many newspapers carry daily law reports, eg *The Times*, *Financial Times*, *Daily Telegraph* and *The Independent*.

The supreme law court is the European Court of Justice, whose decisions are carried in *The Times*.

Secondary sources

A wide range of non-legal sources of information are available, some of which may be quoted in legal situations, however.

1. HSE series of Guidance Notes

Guidance Notes issued by the HSE have no legal status. They are issued on a purely advisory basis to provide guidance on good health and safety practices, specific hazards, etc. There are five series of Guidance Notes – General, Chemical Safety, Plant and Machinery, Medical and Environmental Hygiene.

2. British Standards

These are produced by the British Standards Institute. They provide sound guidance on numerous issues and are frequently referred to by enforcement officers as the correct way of complying with a legal duty.

British Standard 5304 'Safeguarding of machinery' is commonly quoted in conjunction with the duties of employers under the Provision and Use of Work Equipment Regulations 1992 to provide and maintain safe work equipment.

3. Manufacturers' information and instructions

Under Section 6 of the HASWA (as amended by the Consumer Protection Act 1987) manufacturers, designers, importers and installers of 'articles and substances used at work' have a duty to provide information relating to the safe use, storage, etc of their products. Such information may include operating instructions for machinery and plant and hazard data sheets in respect of dangerous substances. Information provided should be sufficiently comprehensive and understandable to enable a judgement to be made on their safe use at work.

4. Safety organisations

Safety organisations, such as the Royal Society for the Prevention of Accidents (RoSPA) and the British Safety Council, provide information in the form of magazines, booklets and videos on a wide range of health and safety-related topics.

5. Professional institutions and trade associations

Many professional institutions, such as the Institute of Occupational Safety and Health (IOSH), the Chartered Institute of Environmental Health (CIEH) and the British Occupational Hygiene Society (BOHS), provide information, both verbally and in written form. Similarly, a wide range of trade associations provide this information to members.

6. Insurance companies
Most insurance companies provide an information service to clients on a wide range of health and safety-related matters.

7. Department of Social Security
This government department publishes annual statistics on claims for industrial injuries benefit and other matters.

8. Published information
This takes the form of textbooks, magazines, law reports, updating services, microfiche systems, films and videos on general and specific topics.

Internal sources
There are many sources of information available within organisations. These include:

1. Existing written information
This may take the form of Statements of Health and Safety Policy, company Health and Safety Codes of Practice, specific company policies, eg on the use and storage of dangerous substances, current agreements with trade unions, company rules and regulations, methods, operating instructions, etc.

Such documentation could be quoted in a court of law as an indication of the organisation's intention to regulate activities in order to ensure legal compliance. Evidence of the use of such information in staff training is essential here.

2. Work study techniques
Included here are the results of activity sampling, surveys, method study, work measurement and process flows.

3. Job descriptions
A job description should incorporate health and safety responsibilities and accountabilities. It should take account of the physical and mental requirements and limitations of certain jobs and any specific risks associated with the job. Representations from operators, supervisors and trade union safety representatives should be taken into account.

Compliance with health and safety requirements is an implied condition of every employment contract, breach of which may result in dismissal or disciplinary action by the employer.

4. Accident and ill-health statistics

Statistical information on past accidents and sickness may identify unsatisfactory trends in operating procedures which can be eliminated at the design stage of safe systems of work.

The use of accident statistics and rates, eg accident incidence rate, as a sole measure of safety performance is not recommended, however, due to the variable levels of accident reporting in work situations. Under-reporting of accidents, common in many organisations, can result in inaccurate comparisons being made between one location and another.

5. Task analysis

Information produced by the analysis of tasks, such as the mental and physical requirements of a task, manual operations involved, skills required, influences on behaviour, hazards specific to the task, and learning methods necessary to impart task knowledge must be taken into account.

Job safety analysis, a development of task analysis, will provide the above information prior to the development of safe systems of work.

6. Direct observation

This is the actual observation of work being carried out. It identifies interrelationships between operators, hazards, dangerous practices and situations and potential risk situations. It is an important source of information in ascertaining whether, for instance, formally designed safe systems of work are being operated or safety practices, imparted as part of former training activities, are being followed.

7. Personal experience

People have their own unique experience of specific tasks and the hazards which those tasks present.

The experiences of accident victims, frequently recorded in accident reports, are an important source of information. Feedback from accidents is crucial in order to prevent repetition of these accidents.

8. Incident recall

This is a technique used in a damage control programme to gain information about near-miss accidents.

9. Product complaints

A record of all product complaints and action taken should be maintained. Such information provides useful feedback in the modification of products and the design of new products. Product complaints may also result in action by the enforcement authorities under Section 6 of the HASAWA and/or civil proceedings in the event of injury, damage or loss sustained as a result of a defective product or defect in a product.

Summary

1. There is an absolute duty on employers to manage health and safety at work.

2. 'Management' is defined as the effective use of resources in the pursuit of organisational goals.

3. Communication is an essential feature of the management process.

4. There is more to successful health and safety management than merely complying with current legislation.

5. Health and safety management covers a wide range of issues, including documentation of procedures and systems, monitoring performance, and the implementation of systems for improving knowledge, attitudes and motivation.

6. In order to be able to manage health and safety effectively, managers must be aware of current information sources.

References

Health and Safety Executive (1991) *Successful Health and Safety Management (HS(G)65)* HMSO, London
Health and Safety Commission (1992) *Management of Health and Safety at Work Regulations 1992 and Approved Code of Practice* HMSO, London
Stranks, J (1994) *Management Systems for Safety* Pitman, London

Chapter 3

Reporting, Recording and Investigation of Accidents, Diseases and Dangerous Occurrences

The reporting and investigation of accidents, occupational diseases and scheduled 'dangerous occurrences', such as a boiler explosion or scaffold collapse, has been a legal requirement for many years. Much of the current legislation is based on the lessons learnt as a result of investigation of these events by the enforcement agencies, government departments and following public enquiries.

All accidents have both indirect and direct causes. They may be associated with unsafe working practices, inadequate supervision and training, human error, poor machinery specification and a host of other causes. (See Figure 4 on page 31.)

Moreover, all accidents, no matter how trivial they may appear, cases of occupational disease, and the type of dangerous occurrence mentioned earlier, represent substantial losses to an organisation. It is essential, therefore, that organisations learn by their mistakes with a view to preventing recurrences of these costly losses.

The law on accident reporting

The Reporting of Injuries, Diseases and Dangerous Occurrences Regulations 1995 (RIDDOR) cover the requirement to *notify and report* certain categories of injury and disease sustained by people at work, together with specified dangerous occurrences and gas incidents to the relevant enforcing authority, ie HSE or local authority. The majority of duties on 'responsibile persons' (as defined) are of an absolute nature.

Principal requirements of RIDDOR
1. Responsible person to *notify* the relevant enforcing authority by the quickest practicable means and subsequently make a *report* within 10 days on the approved form in respect of *death*, any defined *major injury* and *dangerous occurrence* arising out of or in connection with work. (Regulation 3)
2. Duty on an employer to report the *death* of an employee where, as a result of an accident at work, the injured employee dies within one year of the accident. (Regulation 4)

3. Duty on an employer to report cases of *disease* to employees listed in the Schedule. (Regulation 5)
4. Duties on certain persons to report gas incidents. (Regulation 6)
5. Responsible person to keep records of reportable injuries, diseases and dangerous occurrences. (Regulation 7)

Notifiable and reportable major injuries

These are listed in Schedule 1 of RIDDOR, thus:

1. Any fracture, other than to the fingers, thumbs or toes.
2. Any amputation.
3. Dislocation of the shoulder, hip or knee.
4. Loss of sight (whether temporary or permanent).
5. A chemical or hot metal burn to the eye or any penetrating injury to the eye.
6. Any injury resulting from electric shock or electrical burn (including any electrical burn caused by arcing or arcing products) leading to unconsciousness or requiring resuscitation or admittance to hospital for more than 24 hours.
7. Any other injury –
   (a) leading to hypothermia, heat-induced illness or to unconsciousness;
   (b) requiring resuscitation; or
   (c) requiring admittance to hospital for more than 24 hours.
8. Loss of consciousness caused by asphyxia or by exposure to a harmful substance or biological agent.
9. Either of the following conditions which result from the absorption of any substance by inhalation, ingestion or through the skin –
   (a) acute illness requiring medical treatment; or
   (b) loss of consciousness.
10. Acute illness which requires medical treatment where there is reason to believe that this resulted from exposure to a biological agent or its toxins or infected material.

Scheduled dangerous occurrences

A dangerous occurrence is a major incident as a rule which has the potential for significant damage and potential loss of life and which is listed in Schedule 2 of RIDDOR. Dangerous occurrences are classified under five headings.

1. General, eg incidents involving lifting machinery, pressure systems, overhead electric lines.
2. Dangerous occurrences which are reportable in mines, eg fire or ignition of gas, escape of gas, insecure tip.

3. Dangerous occurrences which are reportable in respect of quarries, eg misfires, movement of slopes or faces.
4. Dangerous occurrences which are reportable in respect of relevant transport systems, eg accidents involving any kind of train, incidents at level crossings.
5. Dangerous occurrences which are reportable in respect of an off-shore workplace, eg releases of petroleum hydrocarbon, fire or explosion.

Reportable diseases

These are diseases listed in Schedule 3 of RIDDOR under three classifications.

1. Conditions due to physical agents and the physical demands of work, eg malignant diseases of bones due to ionising radiation, decompression illness.
2. Infections due to biological agents, eg anthrax, brucellosis, leptospirosis.
3. Conditions due to substances, eg poisoning by carbon dispulphide, ethylene oxide and methyl bromide.

Secondly, there are those occupational diseases which are 'prescribed' under the Social Security (Industrial Injuries) (Prescribed Diseases) Regulations 1985, plus various amending Regulations, for the purpose of affected individuals claiming disablement benefit. Typical prescribed diseases include lead and manganese poisoning, pneumoconiosis, viral hepatitis and byssinosis. However, as with reportable diseases, the condition is qualified by the particular occupation, eg miner's nystagmus caused by work in or about a mine. (See Chapter 8)

Reporting procedures under the Regulations

Fatal and major injury accidents, together with scheduled dangerous occurrences, must be:

(a) reported immediately to the enforcing authority by the quickest practicable means, ie telephone, and
(b) reported in writing within 10 days to the enforcing authority on Form 2508.

In the case of an over-three-day injury to a person at work, a written report must be sent to the enforcing authority within seven days of the accident on Form 2508. Occupational diseases listed in Schedule 2 of the Regulations must be reported in the same way following receipt of a written report from a doctor, for example, by medical certificate (statutory sick pay form) on Form 2508A. Similarly, gas incidents must be reported on a Form 2508G.

In all cases, suitable records must be maintained of such reports. This is best done through retaining photocopies of Forms 2508, 2508A and 2508G in a separate file or register.

Accident investigation

There are two principal reasons for investigating all accidents. First, it is essential to ascertain the cause or causes of the accident, which could be associated with poor standards of machinery safety or an unsafe system of work. Secondly, once the causes are identified, it is essential to prevent a recurrence of that accident. Accidents, particularly fatal accidents, can have a serious effect on the morale of the workforce. They may result in lost production, and damage to structural items, plant and machinery, raw materials and finished products, all of which can cause financial losses to the organisation. In particular, there may be a breach of statute, such as the HASAWA, which could result in prosecution of the company by an inspector of the HSE or local authority. Finally, there may be a need for immediate or planned remedial action with a view to securing legal compliance or to prevent further accidents of the same type. In the majority of cases it will be necessary to undertake investigation of an accident to ensure accurate completion of Form 2508 in the case of reportable accidents. However, there is much to be learned from selective investigation of minor injury accidents as these frequently result in some form of loss to the organisation. (See The costs of accidents on page 48.)

While it may not be practicable to investigate all accidents, there are a number of aspects which need consideration prior to an investigation, for instance:

(a) the type of accident, eg machinery, handling goods, use of dangerous substances;

(b) the relative severity of injury, eg fractured leg, minor abrasion;

(c) whether the accident is one of a series of similar accidents, eg people slipping on a particular floor surface, thus identifying a trend in accidents;

(d) whether the accident involved machinery, plant and equipment used at work, or certain substances which could be dangerous, eg strong acids;

(e) the possibility of a breach of the law with subsequent need for a clear statement as to the sequence of events leading to the accident, identification of witnesses, taking of statements from such witnesses, and briefing of defending solicitor in the event of pending prosecution;

(f) the accident, no matter how trivial it may seem, could be the subject of a claim by the injured person on the company's employers' liability insurance arrangements; and

(g) the accident could result in a specific course of action by the trade union representing the injured person.

It is important that management is seen to be acting swiftly, particularly in the event of a fatal or major injury accident. Thorough investigation of an accident can result in a number of management actions, for instance:

(a) the identification of specific training needs for certain groups;

(b) detailed analysis of the task through job safety analysis in order to identify the hazards and precautions necessary, together with certain training requirements;

(c) the issue of specific instructions with regard to systems of work, the use of personal protective equipment, or to improve supervision;

(d) increased employee involvement in health and safety activities, for instance, the establishment of a Health and Safety Committee;

(e) the preparation and issue of a company Code of Practice or Guidance Note dealing with a particular safety procedure;

(f) establishment of a joint management/worker committee or working party to examine a particular trend in accidents and to report back with recommendations;

(g) establishment of a need for better environmental control, eg improved lighting in a specific area or at a particular job;

(h) identification of the need for improved information with regard to the use and storage of potentially dangerous substances; and

(i) clarification as to the particular responsibilities of senior and line management with regard to health and safety generally.

Cases of occupational disease, sometimes referred to as 'slow accidents' because of the long induction period for some diseases, eg 20 years, also merit careful investigation.

The cause–accident–result sequence (see Figure 5) is a useful guide in accident investigation. This sequence shows that there are both direct and indirect causes of accidents but, in many cases, the indirect causes are overlooked in the subsequent investigation. Similarly, there are both direct and indirect results of accidents. The indirect results can be extremely significant, particularly in terms of cost.

'Near misses'

A near miss is an unplanned and unforeseeable incident which could have resulted, but did not result, in death, injury, damage or loss.

INDIRECT CAUSES	DIRECT CAUSES	ACCIDENTS	DIRECT RESULTS	INDIRECT RESULTS
Personal factor *Definition:* Any condition or characteristic of a man that causes or influences him to act unsafely. 1. Knowledge and skill deficiencies: (a) Lack of hazard awareness (b) Lack of job knowledge (c) Lack of job skill. 2. Conflicting motivations: (a) Saving time and effort (b) Avoiding discomfort (c) Attracting attention (d) Asserting independence (e) Seeking group approval (f) Expressing resentment 3. Physical and mental incapacities. **Source causes** *Definition:* Any circumstances that may cause or contribute to the development of an unsafe condition. **Major sources** 1. Production employees 2. Maintenance employees 3. Design and engineering 4. Purchasing practices 5. Normal wear through use 6. Abnormal wear and tear 7. Lack of preventive maintenance 8. Outside contractors.	**Unsafe act** *Definition:* Any act that deviates from a generally recognised safe way of doing a job and increases the likelihood of an accident. **Basic types** 1. Operating without authority 2. Failure to make secure 3. Operating at unsafe speed 4. Failure to warn or signal 5. Nullifying safety devices 6. Using defective equipment 7. Using equipment unsafely 8. Taking unsafe position 9. Repairing or servicing moving or energised equipment 10. Riding hazardous equipment 11. Horseplay 12. Failure to use protection. **Unsafe conditions** *Definition:* Any environmental condition that may cause or contribute to an accident. **Basic types** 1. Inadequate guards and safety devices 2. Inadequate warning systems 3. Fire and explosion hazards 4. Unexpected movement hazards 5. Poor housekeeping 6. Protruding hazards 7. Congestion, close clearance 8. Hazardous atmospheric conditions 9. Hazardous placement or storage 10. Unsafe equipment defects 11. Inadequate illumination, noise 12. Hazardous personal attire.	**The accident** *Definition:* An unexpected occurrence that interrupts work and usually takes this form of an abrupt contact. **Basic types** 1. Struck by 2. Contact by 3. Struck against 4. Contact with 5. Caught in 6. Caught on 7. Caught between 8. Fall to different level 9. Fall on same level 10. Exposure 11. Overexertion/strain.	**Direct results** *Definition:* The immediate results of an accident. **Basic types** 1. 'No results' or near miss 2. Minor injury 3. Major injury 4. Property damage.	**Indirect results** *Definition:* The consequences for all concerned that flow from the direct result of accidents. **For the injured** 1. Loss of earnings 2. Disrupted family life 3. Disrupted personal life 4. And other consequences. **For the company** 1. Injury costs 2. Production loss costs 3. Property damage costs 4. Lowered employee morale 5. Poor reputation 6. Poor customer relations 7. Lost supervisor time 8. Product damage costs.

Figure 5 The cause–accident–result sequence

Studies in the USA by Frank Bird, the exponent of Total Loss Control, showed a correlation between serious or disabling injuries, minor injuries, perhaps requiring first aid treatment only, property damage accidents and accidents with no injury, damage or loss – 'near misses'. These four types of incident were in the ratio 1:10:30:600. In other words, as a result of the studies, for every major or disabling injury, there were 10 minor injuries, 30 property damage accidents and 600 near misses.

What is important is that yesterday's near miss, could be tomorrow's fatal accident. Hence the significance of reporting, recording and investigating such incidents, and the implementation of remedial measures to prevent recurrences.

The costs of accidents

All accidents, cases of occupational disease and scheduled dangerous occurrences represent some degree of loss to a company. There are both direct and indirect costs.

Direct costs

These are sometimes referred to as 'insured costs' and involve the company's liabilities both as an occupier of premises and employer of staff. Companies pay premiums to an insurance company to give them cover against claims made by injured persons. Premiums are determined largely by the past claims history and the risks involved in the business operation. Other direct costs are, perhaps, product liability claims for defective or unsafe products or specific injury claims, which may be settled in or out of a court. Fines imposed by courts for breaches of the law, together with defence costs in such cases, can also be substantial direct costs.

Indirect costs

While many organisations may be fully aware of the direct costs of accidents, very little attention is paid to the indirect costs. Many of these costs may be hidden in other costs and thus not fully recognised, eg production costs, administration costs. Typical indirect costs, many of which can be simply calculated, include the following: treatment costs of the injured employee, eg first aid, transport to hospital, hospital charges, attendance by a local doctor or specialist treatment following the accident; lost time costs, of the injured person, management, first aid staff and others involved; production costs, eg lost production; extra overtime costs to make up production losses; damage costs; and training and supervision costs. It can also be extremely costly to investigate an accident thoroughly in terms of time involvement of management, supervisory staff and witnesses. Other miscellaneous

ACCIDENT COSTS ASSESSMENT

Date: Time: Place of accident:

Details of accident:

Injured person

Name in full:
Address:
Occupation:
Injury details:
Length of service:

Accident costs	£	p
Direct costs		
1. % occupier's liability premium		
2. % increased premiums payable		
3. Claims		
4. Fines and damages awarded in court		
5. Court and legal representation cost		
Indirect costs		
6. *Treatment* First aid		
Transport		
Hospital		
Others		
7. *Lost time* Injured person		
Management		
Supervisor's		
First aider's		
Others		
8. *Production* Lost production		
Overtime payments		
Damage to plant, vehicles, etc		
Training/supervision replacement		
Labour		
9. *Investigation* Management		
Safety adviser		
Others, eg safety representatives		
Liaison with enforcement authority		
10. *Other costs* Replacement of personal items		
— injured person		
— others		
Other miscellaneous costs		
Total Costs		

costs include, perhaps, replacement of damaged personal property and incidental costs incurred by witnesses attending court.

There are no current average costs of accidents. A minor injury accident, perhaps requiring 10 minutes for first aid treatment, may cost very little. On the other hand, if such treatments are frequent, costs can soon mount up. In the case of fatal and major injury accidents, both direct and indirect costs can be substantial, frequently going into six figures.

Accident investigation and costing should, therefore, be undertaken by all organisations. Not only does such an exercise identify causes and costs, but it clearly identifies areas of loss and future loss potential, together with providing feedback on future accident prevention strategies. A sample accident costing form is shown on page 49.

Summary

1. There is a legal duty on owners, occupiers and employers to report certain defined injuries, diseases and dangerous occurrences to the enforcement agencies, namely the HSE or local authority.

2. The investigation of accidents should result in a clear indication of the causes of the accident and of the action necessary to prevent a recurrence.

3. All accidents have both direct and indirect costs to an employer.

4. The cause–accident–result sequence provides a useful guide to accident investigation and potential losses resulting from same.

References

Bird, F E (1974) *Management Guide to Loss Control* Institute Press, Atlanta, Georgia
Health and Safety Executive (1996) *Reporting a Case of Disease* HMSO, London
Health and Safety Executive (1996) *Reporting an Injury or a Dangerous Occurrence* HMSO, London
Morgan, P and Davies, N (1981) The cost of occupational accidents and diseases in Great Britain *Employment Gazette* HMSO, London
Reporting of Injuries, Diseases and Dangerous Occurrences Regulations 1995 (SI 1985 No 2023) HMSO, London

Chapter 4
Principles of Accident Prevention

How often do we hear someone say 'It was an accident! It couldn't be helped.'? Others will take the view that an accident is an act of God over which we have no control. Generally, as far as the accident victim is concerned, accidents have the following characteristics. They are unforeseeable, unintended, unexpected and unplanned.

Accident definitions

A number of definitions of the term 'accident' have been put forward over the years, indicating the differing perceptions and views that exist. The following are some examples.

1. Fenton v Thorley & Co Ltd (1903) AC 443
Some concrete happening which intervenes or obtrudes itself upon the normal course of employment. It has the ordinary everyday meaning of an unlooked-for mishap or an untoward event which is not expected or designed by the victim.

2. Royal Society for the Prevention of Accidents (RoSPA)
An unplanned and uncontrolled event which has led to or could have caused injury to persons, damage to plant or other loss.

3. American Institute of Loss Control
An unintended or unplanned happening that may or may not result in personal injury, property damage, work process stoppage or interference, or any combination of these conditions, under such circumstances that personal injury might have resulted.

4. Little Oxford Dictionary
An event without apparent cause; an unexpected event; an unintentional act; a mishap.

5. Health and Safety Unit, University of Aston in Birmingham
An unexpected, unplanned event in a sequence of events that occurs through a combination of causes. It results in physical harm (injury or disease) to an individual, damage to property, a near miss, a loss, business interruption or any combination of these effects.

It is easy to conclude from these definitions that, in addition to their resulting in many cases from a breach of the law (eg failure to guard machinery properly), most accidents represent some form of loss to an organisation. That loss can be quantified in financial terms, eg increased employer's liability premiums, fines in the courts, business interruption types of loss, damage costs, sickness absence costs and production losses.

Incidents

An incident is a form of accident or situation that does not necessarily result in injury to people. The term 'incident' can be defined as 'an undesired event that could (or does) result in loss', or 'an undesired event that could (or does) downgrade the efficiency of the business operation'. Typical incidents could include, for instance, a minor fire, flooding of a warehouse or an adverse product liability incident involving a company's products, which may result in those products being recalled at very short notice. This has been seen in the past with vehicles, food products and certain electrical appliances. Such incidents result in adverse publicity, recall costs and substantial costs in restoring public confidence in that product.

Certain incidents are classified as 'dangerous occurrences' under RIDDOR, as outlined in Chapter 3.

'Hazard', 'Danger' and 'Risk'

Any strategy in preventing accidents at work must draw a distinction between 'hazard', 'danger' and 'risk'. A 'hazard' is defined as 'the result of a departure from the normal situation, which has the potential to cause death, injury, damage or loss'. 'Danger' is defined as 'liability or exposure to harm; a thing that causes peril'. 'Risk' on the other hand, has a number of definitions – 'a chance of bad consequences', 'exposure to mischance', 'exposure to chance of injury or loss', 'the probability of harm, damage or injury' or 'the probability of a hazard leading to personal injury and the severity of that injury'.

Accidents, therefore, are concerned with two specific aspects – namely the actual danger that exists at a particular point in time and, secondly, how people perceive and measure risk. Typical examples of dangerous situations could include an unfenced floor opening, a badly guarded machine, a slippery floor or, in driving situations, snow on the road. However, it is not until people come on to the scene that an accident can take place. Everyone takes risks in varying degrees, and these may be associated with such factors as individual human traits, training and upbringing. The people-related causes of accidents can, therefore, include over-confidence, ignorance, carelessness, lack of training, apathy and inappropriate attitudes to danger and the risks that can be encountered. The philosophy that 'It couldn't happen to me!' is encountered among many people. (See Chapter 5, Human Factors).

Accident prevention strategies

Strategies aimed at preventing accidents should be geared, firstly, to reducing the objective danger in the workplace ('safe place' strategies) and, secondly, to increasing people's perception of the risks at work ('safe person' strategies). The twin concepts of 'safe place' and 'safe person' must, therefore, feature prominently in any company accident prevention programme with the principal objectives of ensuring compliance with the duties laid on persons at work under HASAWA.

A safe workplace:
'Safe place' strategies are principally concerned with reducing or eliminating the objective dangers which threaten the safety of workers. These include:

1. *Premises:* Premises should be structurally safe in terms of stability, the soundness of floors, staircases and general means of access and egress. The duty to provide safe premises is further reinforced where members of the public may be invited into a premises.

2. *Environment:* Poor standards of working environment are a contributory factor in many accidents. The provision of a sound working environment implies adequate levels of lighting and ventilation, temperature control and the prevention of environmental stressors, such as noise, vibration, dust and fume emission, all of which can affect the health of staff.

3. *Plant and machinery:* Legislation, such as the PUWER (see Chapter 11 for further details), requires plant and machinery to be adequately fenced or controlled in such a way that operators are not exposed to risk of injury. All new machinery and plant should be assessed for hazards prior to acquisition. Maintenance and cleaning systems should take into account the safety requirements of staff engaged in such operations.

4. *Materials:* Materials and substances used at work may be toxic or carcinogenic. They may emit harmful dust or fumes during processing, or represent risks of bodily injury during handling. The duties of manufacturers and suppliers of substances used at work are clearly identified in the HASAWA (Section 6), as amended by the CPA 1987, and the COSHH Regulations.

5. *Processes:* A work process may incorporate a number of machines, materials and differing operating skills. Such factors must be considered during process design and be subject to regular monitoring, particularly with regard to the loading and unloading of machines, the use of potentially dangerous substances in processes, the presence of injurious by-products of manufacture, and the levels of skill and supervision necessary.

6. *Systems of work:* The need for clearly defined and documented safe systems of work is abundantly clear in many work situations. The failure to provide such safe systems, to train operators in their use, to supervise and control those systems and to revise them where necessary, are among the principal causes of industrial accidents. Section 2 of the HASAWA clearly identifies the provision of safe systems of work as an employer's duty.

7. *Supervision and control:* Good standards of safety supervision from the boardroom downwards should be indicated in the employer's Statement of Health and Safety Policy. The specific health and safety related duties of all levels of management and workers should also be clearly identified in job descriptions.

8. *Training:* It is the legal duty of employers to provide information, instruction, training and supervision under the HASAWA. Attention to safety requirements should be stressed during induction training, on-the-job training, and training in specific tasks and operations, such as the operation of lifting equipment and the use of permit to work systems.

Safe persons at work

'Safe person' strategies include the following:

1. *Personal protective equipment:* The provision and use of all forms of personal protective equipment should only be considered as a last resort, ie when all other accident prevention strategies have failed or, alternatively, as an interim measure until one of the 'safe place' strategies mentioned above can be implemented. As an accident prevention strategy, the use of protective equipment relies heavily on the worker wearing the item of personal protection, eg gloves, safety helmet, etc, all the time that he may be exposed to the hazard. Factors such as comfort, choice of the equipment involved, ease of movement, ease of putting the item on and removing it, specific job restrictions created by the equipment, the effects of high temperatures, and ease of cleaning, replacement (as with respirators) and maintenance of parts, are significant when considering the use of personal protective equipment

as a means of protecting workers from hazards. It should be borne in mind that a high level of supervision and control is necessary to ensure constant use of this equipment.

2. *'Vulnerable' groups:* Certain groups of workers, by virtue of their age, physical condition, lack of experience or even general attitude to work, may be more vulnerable to accidents than others. Such groups include young persons, whose experience of hazards could be limited, pregnant women, where there may be specific risks to the unborn foetus, disabled persons, whose physical capacity to undertake certain jobs is considerably reduced, and that very small group of 'accident repeaters', who have the same type of accident regularly. Special consideration is needed in these cases.

3. *'Unsafe behaviour':* Horseplay and other forms of unsafe behaviour can be a feature of some work situations if supervision and control are poor. This has particularly been the case with apprentices in the past. Management must take a very strong line here, with instant dismissal of offenders in extreme cases.

4. *Personal hygiene:* Many substances used in industrial processes can promote occupational skin conditions, in particular, dermatitis. Such substances must be properly controlled, and facilities for maintaining good standards of personal hygiene provided and maintained. This includes washing facilities, which cover showers with hot and cold water, soap, nailbrushes and adequate drying facilities.

5. *Maintaining awareness:* Everyone should be aware of the risks in the workplace. Hazards should be clearly identified in the Statement of Health and Safety Policy, together with the precautions necessary on the part of workers. Methods of increasing and maintaining awareness include the use of posters, training, safety competitions, various forms of safety monitoring, hazard-spotting exercises, hazard-reporting systems and the use of 'Days Lost' notice-boards indicating the number of days lost per month as a result of accidents at work.

Means of preventing accidents

Strategies for preventing accidents take many forms. These include:

1. *Prohibition:* Some processes and practices may be so inherently dangerous that the only way to prevent accidents is by management placing a total prohibition on that activity. This may take the form of a prohibition on the use of a particular substance, such as an identified toxic substance, or of prohibiting people from carrying out unsafe

practices, such as riding on the tines of a fork-lift truck, climbing over moving conveyors or working on roofs without crawlboards.

2. Substitution: The substitution of a less dangerous material or system of work will, in many cases, reduce accident potential. Typical examples are the introduction of remote control handling facilities for direct manual handling operations, the substitution of toluene, a much safer substance, for benzene, and the use of non-asbestos substitutes for boiler and pipe lagging.

3. Change of process: Design or process engineering can usually change a process to ensure better operator protection. Safety aspects of new systems should be considered in the early stages of projects.

4. Process control: This can be achieved through the isolation of a particular process, the use of 'permit to work' systems (see page 57), mechanical or remote control handling systems, restriction of certain operations to highly trained and competent operators, and the introduction of dust and fume arrestment plant.

5. Safe systems of work: Formally designated safe systems of work, with high levels of training, supervision and control, are an important strategy in accident prevention (see below).

6. Personal protective equipment: This entails the provision of items such as safety boots, goggles, aprons, gloves, etc, but is limited in its application as a safety strategy (as discussed on page 54).

Safe systems of work

A safe system of work is defined as 'the integration of men, machinery and materials in the correct environment to provide the safest possible working conditions in a particular working area'.

A safe system of work should incorporate the following features:

(a) a correct sequence of operations;
(b) a safe working area layout;
(c) a controlled environment in terms of temperature, lighting, ventilation, dust control, humidity control, sound pressure levels and radiation hazards; and
(d) clear specification of safe practices and procedures for the task in question.

Safe systems of work are generally designed through the technique of 'job safety analysis'.

Job safety analysis

Job safety analysis is based on task analysis and takes ⟍
stages. The first stage entails an examination of the task ⟍
attention being paid to the following points: the job ⟍
operations; the details of machinery, equipment, materi_____ substances used; the hazards; the degree of risk, ie low, medium or high
risk; the specific tasks; the work organisation; and the form of operator
protection necessary. Each of these points is detailed on a standard
form. The second stage of the operation assesses:

(a) the specific job operations;
(b) the hazards associated with these operations, eg risk of hand
    injury;
(c) the skills required to perform the task safely in terms of knowl-
    edge and individual behaviour;
(d) the external influences on behaviour in terms of:
    (i)   the nature of the influence, eg low temperatures;
    (ii) the source of the influence, eg refrigeration system; and
    (iii) the activities involved, eg stacking of refrigerated food
          products;
(e) the learning method, eg induction training in the use of low
    temperature protective wear, safe stacking, and the use of the
    alarm system.

Many organisations now produce safe systems of work on a flowchart
basis, as shown in the example in Figure 6, copies of which are
displayed in the operational area and used in staff training activities.

Permit to work systems

Such a system is a form of safe system of work. It is operated
fundamentally where there is a high degree of foreseeable risk, eg entry
into confined spaces or vessels containing agitators. A typical permit to
work system incorporates a number of clearly defined stages:

(a) Assessment: This stage entails an assessment of the work to be done,
the method, materials and equipment to be used, and the inherent
hazards. Such an assessment must be made by a senior member of
the management team with a view to identifying the safest possible
ways to undertake a potentially dangerous task or operation.

(b) Withdrawal from service: The next stage entails withdrawal of the plant
from service, designation of this fact by warning signs or fencing
around the area, limitation of access in certain cases and restriction
of entry to identified personnel.

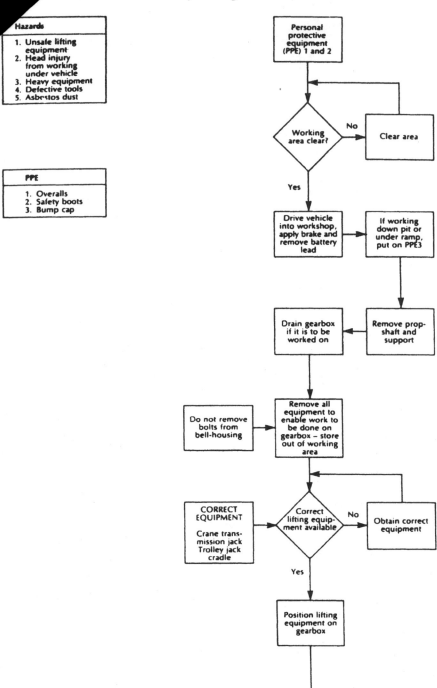

Hazards

1. Unsafe lifting equipment
2. Head injury from working under vehicle
3. Heavy equipment
4. Defective tools
5. Asbestos dust

PPE

1. Overalls
2. Safety boots
3. Bump cap

Personal protective equipment (PPE) 1 and 2

Working area clear? — No → Clear area

Yes

Drive vehicle into workshop, apply brake and remove battery lead → If working down pit or under ramp, put on PPE3

Drain gearbox if it is to be worked on ← Remove prop-shaft and support

Do not remove bolts from bell-housing → Remove all equipment to enable work to be done on gearbox – store out of working area

CORRECT EQUIPMENT

Crane transmission jack
Trolley jack
cradle

Correct lifting equipment available — No → Obtain correct equipment

Yes

Position lifting equipment on gearbox

Figure 6   Safe system of work

58

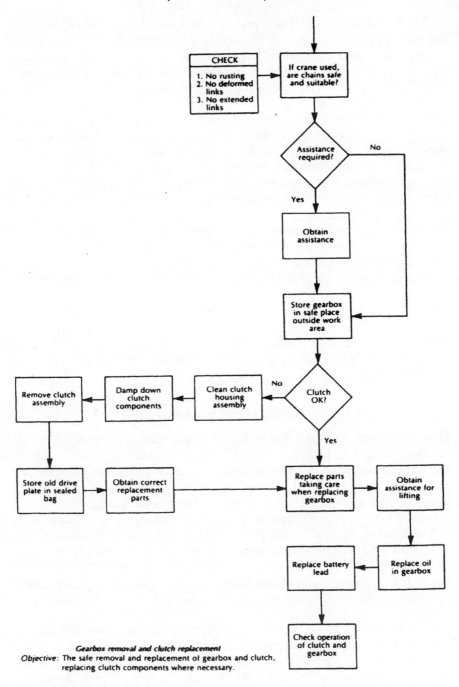

Gearbox removal and clutch replacement
Objective: The safe removal and replacement of gearbox and clutch,
replacing clutch components where necessary.

(c) Isolation: Physical, electrical and/or mechanical isolation of the plant implies controlled restriction of access and the physical locking off of sources of power. In certain cases it may be necessary to undertake environmental testing to ascertain whether the use of breathing apparatus is necessary before entry into a confined space.

(d) Completion of work and return to service: On completion of the scheduled operation, details of which must be clearly identified in the permit to work, the certificate is cancelled and returned to the originator for checking to ensure work has been satisfactorily completed. The plant is then handed back for normal use to the manager responsible for same who, in turn, should check the work has been completed satisfactorily.

All the above headings feature in the permit to work form and, most importantly, require a signature certifying completion of the particular stage before the next stage can proceed.

Permits to work are generally produced in triplicate, with the original going to the person undertaking the work, the first copy to the departmental manager in whose area the work is being undertaken and the second copy being retained by the originator. At completion of the operation and final clearance of the permit, all copies are returned to the originator. It is standard practice to maintain a record of all permits to work issued.

Competent persons

One way of ensuring the operation of a safe system of work is by the designation and employment of specifically trained operators or external contractors who appreciate the risks involved. This may be for undertaking certain inspections, eg of abrasive wheels and power presses, issuing permits to work and supervising their operation, or for undertaking work where there is a high degree of foreseeable risk.

The expression 'competent person' occurs frequently in construction safety law and is increasingly appearing in new legislation, such as the Noise at Work Regulations 1989. Under the Construction (Health, Safety and Welfare) Regulations 1996 certain inspections, examinations, operations and supervisory duties must be undertaken by competent persons.

Generally, the onus is on the employer to decide whether persons are competent to undertake certain duties. An employer might do this by reference to the person's training, qualifications and experience. Broadly, a competent person should have practical and theoretical knowledge as well as sufficient experience of the particular machinery, plant or procedure involved to enable him to identify defects or weaknesses during plant and machinery examinations, and to assess their importance in relation to the strength and function of that plant and machinery. [Brazier v Skipton Rock Company Ltd (1962) 1 AER 955]

Competent persons are involved in a wide range of activities under specific legislation, and their role and function is increasingly being specified in ACOPs supporting particular Regulations.

Such a person should be authorised in writing to undertake specified duties, and should be subject to a practical test and oral examination at least by a senior supervisor prior to any declaration regarding his suitability to undertake such duties.

See also 'Competent persons' under the MHSWR on page 20.

Safety monitoring systems

Safety monitoring is concerned with the measurement and evaluation of safety performance. It may take the following forms.

1. Safety surveys: This is a detailed examination of a number of critical areas of operation or, perhaps, an in-depth study of all health and safety related activities in a workplace.

2. Safety tours: These are an unscheduled examination of a working area, frequently undertaken as a group exercise (eg foreman, safety representative and safety committee member), to assess general compliance with safety requirements (eg fire protection measures and use of machinery safety devices).

3. Safety audits: A safety audit fundamentally subjects each area of an organisation's activities to a systematic critical examination with the object of minimising injury and loss. It generally takes the form of a series of questions directed to examining factors such as the operation of safe systems of work, compliance with the Statement of Health and Safety Policy and the operation of hazard reporting systems. A specimen audit form can be seen on pages 62–6.

4. Safety inspections: A scheduled inspection of a premises or working area to assess levels of legal compliance and observation of company safety procedures. Safety inspections are frequently undertaken by company safety specialists and trade union safety representatives.

5. Safety sampling: A system designed to measure by random sampling the accident potential in a workplace or process by identifying defects in safety performance or omissions. Observers follow a prescribed route through the working area noting deficiencies in performance, eg concerning the wearing of personal protective equipment or the use of correct manual handling techniques. In some cases, individual topics in the safety sampling exercise are ranked according to importance with a maximum number of points achievable. At the end of the exercise a total score is identified which gives an indication of the performance level at that point in time. A typical safety sampling exercise is shown in Figure 7, Health and safety review.

6. Hazard and operability studies: Such studies incorporate the application of formal critical examination to the process and engineering intentions regarding new facilities. The principal aim of such a study is to assess the hazard potential arising from the incorrect operation of equipment and the consequential effects on the facility. Such an operation enables remedial action to be taken at a very early stage.

7. Damage control: Levels of damage are an indication of future accident potential. Damage control operates on the philosophy that non-injury accidents are just as important as injury accidents. The elimination of the causes of accidents resulting in damage to property, plant and products frequently results in a reduction in injury accidents.

Specimen Safety Audit

	YES	NO
1. DOCUMENTATION		

1. Are management aware of all health and safety legislation applying to their workplace?
   Is this legislation available to management and employees?
2. Have all Approved Codes of Practice, HSE Guidance Notes and internal codes of practice been studied by management with a view to ensuring compliance?
3. Does the existing Statement of Health and Safety Policy meet current conditions in the workplace?
   Is there a named manager with overall responsibility for health and safety?
   Are the 'organisation and arrangements' to implement the Health and Safety Policy still adequate?
   Have the hazards and precautions necessary on the part of staff and other persons been identified and recorded?
   Are individual responsibilities for health and safety clearly detailed in the Statement?
4. Do all job descriptions adequately describe individual health and safety responsibilities and accountabilities?
5. Do written safety guidelines exist for all potentially hazardous operations?
   Is Permit to Work documentation available?

6. Has a suitable and sufficient assessment of the risks to staff and other persons been made, recorded and brought to the attention of staff and other persons?
   Have other risk assessments in respect of:

   (a) substances hazardous to health;
   (b) risks to hearing;
   (c) work equipment;
   (d) personal protective equipment;
   (e) manual handling operations; and
   (f) display screen equipment,

   been made, recorded and brought to the attention of staff and other persons?
7. Is the fire certificate available and up to date?
   Is there a record of inspections of the means of escape in the event of fire, fire appliances, fire alarms, warning notices, fire and smoke detection equipment?
8. Is there a record of inspections and maintenance of work equipment, including guards and safety devices?
   Are all examination and test certificates available, eg lifting appliances and pressure systems?
9. Are all necessary licences available, eg to store petroleum spirit?
10. Are workplace health and safety rules and procedures available, promoted and enforced?
    Have these rules and procedures been documented in a way which is comprehensible to staff and others, eg Health and Safety Handbook?
    Are disciplinary procedures for unsafe behaviour clearly documented and known to staff and other persons?
11. Is a formally written emergency procedure available?
12. Is documentation available for the recording of injuries, near misses, damage only accidents, diseases and dangerous occurrences?
13. Are health and safety training records maintained?
14. Are there documented procedures for regulating the activities of contractors, visitors and other persons working on the site?
15. Is hazard reporting documentation available to staff and other persons?
16. Is there a documented planned maintenance system?
17. Are there written cleaning schedules?

2. HEALTH AND SAFETY SYSTEMS

| | YES | NO |

1. Have competent persons been appointed to:

   (a) co-ordinate health and safety measures; and
   (b) implement the emergency procedure?

   Have these persons been adequately trained on the basis of identified and assessed risks?

Is the role, function, responsibility and accountability of competent persons clearly identified?

2. Are there arrangements for specific forms of safety monitoring, eg safety inspections, safety sampling?
   Is a system in operation for measuring and monitoring individual management performance on health and safety issues?

3. Are systems established for the formal investigation of accidents, ill health, near misses and dangerous occurrences?
   Do investigation procedures produce results which can be used to prevent future incidents?
   Are the causes of accidents, ill health, near misses and dangerous occurrences analysed in terms of the failure of established safe systems of work?

4. Is a hazard reporting system in operation?

5. Is a system for controlling damage to structural items, machinery, vehicles, etc in operation?

6. Is the system for joint consultation with trade union safety representatives and staff effective?
   Are the role, constitution and objectives of the Health and Safety Committee clearly identified?
   Are the procedures for appointing or electing committee members and trade union safety representatives clearly identified?
   Are the available facilities, including training arrangements, known to committee members and trade union safety representatives?

7. Are the capabilities of employees as regards health and safety taken into account when entrusting them with tasks?

8. Is the provision of first aid arrangements adequate?
   Are first aid personnel adequately trained and retrained?

9. Are the procedures covering sickness absence known to staff?
   Is there a procedure for controlling sickness absence?
   Are managers aware of the current sickness absence rate?

10. Do current arrangements ensure that health and safety implications are considered at the design stage of projects?

11. Is there a formal annual health and safety budget?

3. PREVENTION AND CONTROL PROCEDURES

	YES	NO

1. Are formal inspections of machinery, plant, hand tools, access equipment, electrical equipment, storage equipment, warning systems, first aid boxes, resuscitation equipment, welfare amenity areas, etc undertaken?
   Are machinery guards and safety devices examined on a regular basis?

2. Is a Permit to Work system operated where there is a high degree of foreseeable risk?

3. Are fire and emergency procedures practised on a regular basis?
   Where specific fire hazards have been identified, are they catered for in the current fire protection arrangements?

Are all items of fire protection equipment and alarms tested, examined and maintained on a regular basis?

Are all fire exits and escape routes marked, kept free from obstruction and operational?

Are all fire appliances correctly labelled, sited and maintained?

4. Is a planned maintenance system in operation?
5. Are the requirements of cleaning schedules monitored?

   Is housekeeping of a high standard, eg material storage, waste disposal, removal of spillages?

   Are all gangways, stairways, fire exits, access and egress points to the workplace maintained and kept clear?
6. Is environmental monitoring of temperature, lighting, ventilation, humidity, radiation, noise and vibration undertaken on a regular basis?
7. Is health surveillance of persons exposed to assessed health risks undertaken on a regular basis?
8. Is monitoring of personal exposure to assessed health risks undertaken on a regular basis?
9. Are local exhaust ventilation systems examined, tested and maintained on a regular basis?
10. Are arrangements for the storage and handling of substances hazardous to health adequate?

    Are all substances hazardous to health identified and correctly labelled, including transfer containers?
11. the appropriate personal protective equipment available?

    Is the personal protective equipment worn or used by staff consistently when exposed to risks?

    Are storage facilities for items of personal protective equipment provided?
12. Are welfare amenity provisions, ie sanitation, hand washing, showers and clothing storage arrangements adequate?

    Do welfare amenity provisions promote appropriate levels of personal hygiene?

4. INFORMATION, INSTRUCTION, TRAINING AND SUPERVISION

	YES	NO

1. Is the information provided by manufacturers and suppliers of articles and substances for use at work adequate?

   Do employees and other persons have access to this information?
2. Is the means of promoting health and safety adequate?

   Is effective use made of safety propaganda, eg posters?
3. Do safety signs meet the requirements of the Safety Signs Regulations 1980 and Health and Safety (Safety Signs and Signals) Regulations 1996?

   Are safety signs adequate in terms of the assessed risks?
4. Are fire instructions prominently displayed?
5. Are hazard warning systems adequate?
6. Are the individual training needs of staff and other persons assessed on a regular basis?

7. Is staff health and safety training undertaken:

   (a) at the induction stage;
   (b) on employees being exposed to new or increased risks
       because of:
       (i)   transfer or change in responsibilities;
       (ii) the introduction of new work equipment or a change
             respecting existing work equipment;
       (iii) the introduction of new technology;
       (iv) the introduction of a new system of work or change in
             an existing system of work?

   Is the above training:

   (a) repeated periodically;
   (b) adapted to take account of new or changed risks; and
   (c) carried out during working hours?
8. Is specific training carried out regularly for first aid staff, fork-
   lift truck drivers, crane drivers and others exposed to specific
   risks?
   Are selected staff trained in the correct use of fire appliances?

5. FINAL QUESTION

	YES	NO

Are you satisfied that your organisation is as safe and healthy as
you can reasonably make it, or that you know what action must
be taken to achieve that state?

ACTION PLAN

1. Immediate action

2. Short-term action (14 days)

3. Medium-term action (6 months)

4. Long-term action (2 years)

_____Auditor   Date_____

Unit ....

Date ....

WORKPLACE	MAX	RATING	SAFETY SYSTEMS	MAX	RATING
Cleaning & housekeeping	10		Policy Statement & plan	10	
Structural safety	10		Safe systems of work	20	
Machinery safety	20		Accident reporting/costing	10	
Fire protection	20		Sickness absence reporting/ costing	10	
Chemical safety	10		Hazard reporting	10	
Electrical safety	10		Cleaning schedules	10	
Internal transport	10		Catering hygiene & safety	10	
Access equipment	5		Emergency procedure	10	
Internal storage	5		Safety monitoring procedure	10	
MAX	100		MAX	100	
PEOPLE			ENVIRONMENT		
Personal protection	20		Environmental control – internal	10	
Chemical handling	10		Environmental control – external	5	
Manual handling	10		Noise control – internal	10	
Propaganda	20		Noise control – external	5	
Training	20		Welfare amenity provisions	20	
Safe behaviour	20				
MAX	100		MAX	50	
			TOTAL SCORE MAX	350	

Figure 7   Health and safety review

Accident and ill-health rates

The following rates are used in the calculation of accident and ill health rates:

$$\text{Frequency Rate} = \frac{\text{Total number of accidents}}{\text{Total number of man-hours worked}} \times 100,000$$

$$\text{Incident Rate} = \frac{\text{Total number of accidents}}{\text{Number of persons employed}} \times 1,000$$

$$\text{Severity Rate} = \frac{\text{Total number of days lost}}{\text{Total number of man-hours worked}} \times 1,000$$

$$\text{Mean Duration Rate} = \frac{\text{Total number of days lost}}{\text{Total number of accidents}}$$

$$\text{Duration rate} = \frac{\text{Number of man-hours worked}}{\text{Total number of accidents}}$$

$$\text{Sickness Absence Rate} = \frac{\text{Total man-days lost through sickness}}{\text{Total man-days worked}} \times 100$$

Risk assessment

A risk assessment may be defined as: 'an identification of the hazards present in an undertaking and an estimate of the extent of the risks involved, taking into account whatever precautions are already being taken'.

It is essentially a four-stage process:

(a) identification of all the hazards;
(b) measurement of the risks;
(c) evaluation of the risks; and
(d) implementation of measures to eliminate or control the risks.

There are different approaches which can be adopted in the workplace, eg:

(a) examination of each activity which could cause injury;
(b) examination of hazards and risks in groups, eg machinery, substances, transport; and/or
(c) examination of specific departments, sections, offices, construction sites.

In order to be suitable and sufficient and to comply with legal requirements, a risk assessment must:

(a) identify all the hazards associated with the operation, and evaluate the risks arising from those hazards, taking into account current legal requirements;

(b) record the significant findings if more than five persons are employed, even if located in different locations;

(c) identify any group of employees, or single employees as the case may be, who are especially at risk;

(d) identify others who may be specially at risk, eg visitors, contractors, members of the public;

(e) evaluate existing controls, stating whether or not they are satisfactory and, if not, what action should be taken;

(f) evaluate the need for information, instruction, training and supervision;

(g) judge and record the likelihood of an accident occurring as a result of uncontrolled risk, including the 'worst case' likely outcome;

(h) record any circumstances arising from the assessment where serious and imminent danger could arise; and

(i) provide an action plan giving information on implementation of additional controls, in order of priority, and with a realistic timescale.

Recording the assessment

The assessment must be recorded in organisations where more than five persons are employed. (Electronic methods of recording are acceptable.) The assessment must incorporate details of items (a) to (i) above.

Generic assessments

These are assessments produced once only for a given activity or type of workplace. In cases where an organisation has several locations, or in situations where the same activity is undertaken, a generic risk assessment could be carried out for a specific activity to cover all locations. Similarly, where operators work away from the main location and undertake a specific task, eg installation of telephones or servicing of equipment, a generic assessment should be produced.

For generic assessments to be effective,

(a) 'worst case' situations must be considered; and

(b) provision should be made for monitoring implementation of assessment controls which are/are not relevant at a particular location, and for determining what action needs to be taken to implement the relevant tasks outlined in the assessment.

In certain cases, there may be risks which are specific to one situation only, and these risks may need to be incorporated in a separate part of the generic risk assessment.

Maintaining the risk assessment

The risk assessment must be *maintained*. This means that any significant change to a workplace, process or activity, or the introduction of any new process, activity or operation, should be subject to risk assessment. If new hazards come to light, then these should also be subject to risk assessment. The risk assessment, furthermore, should be periodically reviewed and updated.

This is best achieved by a suitable combination of safety inspection and monitoring techniques, which require corrective and/or additional action where the need is identified. The process for monitoring, review and corrective action for risk assessment is shown on pages 69–70.

Typical monitoring systems include:

(a) preventive maintenance inspections;
(b) safety representative/committee inspections;
(c) statutory and maintenance scheme inspections, tests and examinations;
(d) safety tours and inspections;
(e) occupational health surveys;
(f) air monitoring; and
(g) safety audits.

Useful information on checking performance against control standards can also be obtained reactively from the following activities:

(a) accident and ill-health investigation;
(b) investigation of damage to plant, equipment and vehicles; and
(c) investigation of 'near miss' situations.

Reviewing the risk assessment

The frequency of review depends upon the level of risk in the operation, and should not normally exceed 10 years. Further, a review is necessary if a serious accident occurs in the organisation or elsewhere; or a check on the risk assessment shows a gap in assessment procedures.

Risk/hazard control

Once the risk or hazard has been identified and assessed, employers must either prevent the risk arising or, alternatively, control same. Much will depend upon the magnitude of the risk in terms of the controls applied. In certain cases, the level of competence of operators may need to be assessed prior to their undertaking certain work, eg work on electrical systems.

A typical hierarchy of control, from high-risk to low-risk, is indicated below.

1. **Elimination** of the risk completely, eg prohibiting a certain practice or the use of a certain hazardous substance.
2. **Substitution** with something less hazardous or risky.
3. **Enclosure** of the risk in such a way that access is denied
4. **Fitting guards or installing safety devices** to prevent access to danger points or zones on work equipment and machinery.
5. **Safe systems of work** that reduce the risk to an acceptable level.
6. **Written procedures**, eg job safety instructions, that are known and understood by those affected.
7. **Adequate supervision**, particularly in the case of young or inexperienced persons.
8. **Training** of staff to appreciate the risks and hazards.
9. **Information**, eg safety signs, warning notices.
10. **Personal protective equipment**, eg eye, hand, head and other forms of body protection.

In many cases, a combination of the above control methods may be necessary.

It should be appreciated that the amount of management control necessary will increase proportionately for the controls lower down this list. In other words, item 1 indicates no management control is needed, whereas item 10 requires a high degree of control.

The role of the health and safety practitioner

Practitioners in occupational health and safety include occupational health nurses, health and safety managers/advisers, safety officers, occupational physicians, occupational hygienists and safety representatives/stewards. Their various roles are outlined below.

Occupational health nurses

Generally an occupational health nurse would be a qualified nurse, eg RGN, with a separate qualification in occupational health nursing, for instance the Occupational Health Nursing Certificate (OHNC).

The occupational health nurse's role consists of eight main elements:

(a) health supervision;
(b) health education;
(c) environmental monitoring and occupational safety;
(d) counselling;

(e) treatment services;
(f) rehabilitation and resettlement;
(g) unit administration and record systems; and
(h) liaison with other agencies, eg employment medical advisers of the Employment Medical Advisory Service.

The Royal College of Nursing lists the following duties which could be undertaken by a fully trained occupational health nurse:

(a) health assessment in relation to the individual worker and the job to be performed;
(b) noting normal standards of health and fitness and any departures or variations from these standards;
(c) referring to the occupational physician or doctor such cases which, in the opinion of the nurse, require further investigation and medical, as distinct from nursing, assessment;
(d) health supervision of vulnerable groups, eg young persons;
(e) routine visits to and surveys of the working environment, and informing as necessary the appropriate expert when a particular problem requires further specialised investigation;
(f) employee health counselling;
(g) health education activities in relation to groups of workers;
(h) the assessment of injuries or illness occurring at work and treatment or referral as appropriate;
(i) responsibility for the organisation and administration of occupational health services, and the control and safe-keeping of non-statutory personal health records; and
(j) a teaching role in respect of the training of first aid personnel and the organisation of emergency services.

Health and safety managers/advisers and safety officers

The increase in attention to safety at work following the HASAWA has resulted, to some extent, in the development, particularly in large multi-site organisations, of a two-tier system of specialists in occupational health and safety. The larger companies tend to promote this two-tier system with a health and safety manager, frequently reporting at director or board level, supported by local safety officers who operate in a specific work location. Health and safety managers/advisers are concerned with advising the organisation on policy issues and with ensuring the integration of health and safety considerations into the normal operating procedures of the organisation. They may come from a number of backgrounds, eg HSE, or from particular disciplines, such as chemistry, engineering, etc.

The health and safety manager would formulate policy on matters relating to, for instance, the development of local Statements of Health and Safety Policy, health and safety training, promotion of health and safety related issues, control of the working environment, disaster planning and control, and accident reporting procedures. In some cases he may be responsible for monitoring the performance of full-time or part-time safety officers in individual units. His objectives are principally concerned with reducing losses, improving management performance and generally raising the profile of health and safety in the organisation.

Safety officers may be appointed on a full-time or part-time basis. Their role and function are specifically related to the workplace in which they operate. While there is no current legal requirement for such persons to be trained, increasingly companies are seeking membership of the Institution of Occupational Safety and Health (IOSH) as a basic qualification for this task. Training courses at Certificate and Diploma level are organised through the National Examination Board in Occupational Safety and Health (NEBOSH), and such courses can be taken on a full-time, part-time or block-release basis at local technical colleges, through the Royal Society for the Prevention of Accidents (RoSPA) and other approved training providers.

IOSH has indicated the following as typical duties of a safety officer:

(a) advising line management in order to assist it to fulfil its responsibility for safety, including:
    (i) advising on safety aspects in the design and use of plant and equipment and the checking of new equipment before commissioning; and
    (ii) carrying out periodic inspections to identify unsafe plant, unsafe working conditions and unsafe practices, to report on the results of such inspections and to make recommendations for remedying any defects found;

(b) advising upon the drawing up and implementation of safe systems of work, and the provision and use of personal protective equipment;

(c) advising on legal requirements affecting health and safety;

(d) participating in the work of safety committees and joint consultations affecting the workforce;

(e) promoting and, where appropriate, participating in safety education programmes to assist in safety consciousness at all levels within the organisation and specifically to teach supervisors to develop safe working conditions;

(f) working in collaboration with the training department, where this exists, to secure regular safety training of employees;

(g) providing information about accident prevention techniques and preparing visual aids, including posters, slides, film strips, etc for safety training purposes;

(h) assessing possible causes of injury and circumstances likely to produce an accident, the compilation of necessary reports, and tendering advice to prevent recurrence;

(i) recording of accident statistics and presenting information in appropriate form for the use of management and others in ensuring safety performance;

(j) maintaining liaison with other departments, including medical and training departments, with official bodies such as government inspectorates, local authorities and fire authorities, and with outside bodies such as RoSPA and the Fire Protection Association; and

(k) keeping up to date with modern processes and techniques, with special reference to safety.

Occupational physicians

An occupational physician is a registered medical practitioner. He should preferably hold a recognised qualification in occupational medicine, such as membership of the Faculty of Occupational Medicine or other appropriate organisation. The British Medical Association has identified the role of the occupational physician as encompassing the following:

1. The effect of health on the capacity to work, which includes:
   (a) Provision of advice to employees on all health matters relating to their working capacity;
   (b) Examination of applicants for employment and advice as to their placement;
   (c) Immediate treatment of surgical and medical emergencies occurring at the place of employment;
   (d) Examination and continued observation of persons returning to work after absence due to illness or accident and advice on suitable work;
   (e) Health supervision of all employees with special reference to young persons, pregnant women, elderly persons and disabled persons.
2. The effect of work on health, which includes:
   (a) Responsibility for nursing and first aid services;
   (b) Study of the work and working environment and how they affect the health of employees;
   (c) Periodical examination of employees exposed to special hazards in respect of their employment;
   (d) Advising management regarding:
       (i) the working environment in relation to health;
       (ii) occurrence and significance of hazards;
       (iii) accident prevention; and
       (iv) statutory requirements in relation to health;
   (e) Medical supervision of the health and hygiene of staff and

facilities, with particular reference to canteens, kitchens, etc and those working in the production of foods or drugs for sale to the public;

(f) Arranging and carrying out such education work in respect of the health, fitness and hygiene of employees as may be desirable and practicable;

(g) Advising those committees within the organisation which are responsible for the health, safety and welfare of employees.

Occupational hygienists

'Occupational hygiene' has been defined as being concerned with the identification, measurement, evaluation and control of environmental factors, such as noise and radiation, and contaminants arising from work operations which may adversely affect the health of employees.

Entry to the profession is controlled by the British Examination and Registration Board in Occupational Hygiene (BERBOH). Occupational hygienists frequently operate on a consultancy basis, undertaking measurement and evaluation of airborne hazards, such as asbestos, fumes, dusts and vapours, in addition to physical phenomena, such as noise, vibration and radiation. The significance of occupational hygiene as a specialist role in occupational health and safety has increased with the advent of the COSHH Regulations.

Summary

1. All accidents and incidents represent varying degrees of loss to organisations.

2. Accident prevention strategies should take into account the concepts of 'safe place' and 'safe person', with the emphasis being on 'safe place' strategies.

3. Safe systems of work are the basis for effective levels of safety management.

4. All organisations should operate one or more forms of safety monitoring as a means of establishing standards and measuring performance against these standards.

5. When considering the appointment of health and safety practitioners, reference should be made to the accepted standards of competence and professional qualifications outlined in this chapter.

References

Atherley, G R C (1978) *Occupational Health and Safety Concepts* Applied Science Publishers Ltd, London

Bird, F E and Loftus, R G (1984) *Loss Control Management* RoSPA, Birmingham

Chemical Industries Association Ltd (1975) *Safety Audits* Chemical Industries Association, London

Department of Employment (1974) *Accidents in Factories: The Pattern of Causation and Scope for Prevention* HMSO, London

Health and Safety Executive (1982) *Guidelines for occupational health services Guidance Note HS(G) 20* HMSO, London

Health and Safety Executive (1992) *Personal Protective Equipment at Work Regulations 1992 and Guidance on Regulations* HMSO, London

Health and Safety Commission (1992) *Management of Health and Safety at Work Regulations 1992 and Approved Code of Practice* HMSO, London

Stranks, J (1997) *Handbook of Health and Safety Practice* Pitman, London

Stranks, J (1996) *The Law and Practice of Risk Assessment* Pitman, London

Stranks, J (1995) *Management Systems for Safety* Pitman, London

Conclusion to Part I

Part I has examined the management and administration aspects of occupational health and safety. Of particular significance are the duties of employers and employees under the HASAWA and the importance of clearly written Statements of Health and Safety Policy.

The Policy Statement forms the framework for all health and safety related activities in the organisation. In particular, it identifies the duties and responsibilities of all levels of management and employees, the organisation and arrangements for ensuring implementation of the Policy and systems for measuring performance.

However, the legal requirements should be considered the minimum acceptable standard. There should now be a move away from the purely legalistic approach to health and safety towards a philosophy of 'asset protection'. Without people, companies would not be in business; the loyal, skilled, long-serving employee is any company's most valuable asset. The maintenance of good standards of safety, health and welfare is one of the most effective ways of ensuring continuity of that loyalty.

Frank Bird, the American exponent of Total Loss Control, stated many years ago that 'Accidents downgrade the system'. Apart from resulting in substantial losses to a company, they result in reduced morale, poor industrial relations, loss of company image in the market-place and loss of public confidence. The need, therefore, for clearly developed safe systems of work, accident reporting procedures and well-developed safety monitoring systems cannot be over-emphasised. Such operations should form part of the company 'culture' and not be viewed in isolation as is so frequently the case.

Health and safety practitioners have a significant contribution to make in reducing the losses. However, they must be properly trained and set clearly defined objectives which are capable of measurement.

Part 2
People at Work

Chapter 5
Human Factors

The MHSWR 1992 brought in the need for employers to place greater emphasis on people, their capabilities and fallibilities. Regulation 11 states that 'every employer shall, in entrusting tasks to his employees, take into account their capabilities as regards health and safety.'

The ACOP qualifies this requirement as follows: 'When allocating work to employees, employers should ensure that the demands of the job do not exceed the employees' ability to carry out the work without risk to themselves or others. Employers should take account of the employees' capabilities and the level of their training, knowledge and experience. If additional training is needed, it should be provided.'

The HSE publication 'Human factors in industrial safety' [HS(G)48] defines 'human factors' as a range of issues including the perceptual, physical and mental capabilities of people and the interactions of individuals with their job and the working environments, the influence of equipment and system design on human performance and, above all, the organisational characteristics which influence safety related behaviour at work.

There are three areas of influence on people at work, namely:

(a) the organisation;
(b) the job; and
(c) personal factors.

These are directly affected by:

(a) the system for communication within the organisation; and
(b) the training systems and procedures in operation;

all of which are directed at preventing human error.

The organisation
Characteristics of organisations which influence safety-related behaviour include:

(a) the need to promote a positive climate in which health and safety is seen by both management and employees as being fundamental to the organisation's day-to-day operations ie they must create a positive safety culture;

(b) the need to ensure that policies and systems which are devised for the control of risk from the organisation's operations take proper account of human capabilities and fallibilities;

(c) commitment to the achievement of progressively higher standards which is shown at the top of the organisation and cascaded through successive levels of same;

(d) demonstration by senior management of their active involvement, thereby galvanising managers throughout the organisation into action; and

(e) leadership whereby an environment is created which encourages safe behaviour.

The job

Successful management of human factors and the control of risk involves the development of systems designed to take proper account of human capabilities and fallibilities. Using techniques like job safety analysis, jobs should be designed in accordance with ergonomic principles (see Chapter 6) so as to take into account limitations in human performance.

Major considerations in job design include:

(a) identification and comprehensive analysis of the critical tasks expected of individuals and appraisal of likely errors;

(b) evaluation of required operator decision making and the optimum balance between human and automatic contributions to safety actions;

(c) application of ergonomic principles in the design of man–machine interfaces, including displays of plant and process information, control devices and panel layouts;

(d) design and presentation of procedures and operating instructions;

(e) organisation and control of the working environment including workspace, access for maintenance, noise, lighting and thermal conditions;

(f) provision of correct tools and equipment;

(g) scheduling of work patterns, including shift organisation, control of fatigue and stress, and arrangements for emergency operations; and

(h) efficient communications, both immediate and over periods of time.

Personal factors

This aspect is concerned with how factors such as attitude, motivation, training, human error and the perceptual, physical and mental capabilities of people can interact with health and safety issues.

Attitudes are directly connected with an individual's self-image, the influence of groups and compliance with group norms or standards and, to some extent, opinions, including superstitions, like 'all accidents are acts of God!' Changing attitudes is difficult. They may be affected by past experience, the level of intelligence, specific motivation, financial gain and by the skills available to the individual. There is no doubt, however, that management example is the strongest of all motivators for bringing about attitude change.

Important factors in motivating people to work safely include joint consultation in planning the work organisation, the use of working parties to define objectives, attitudes currently held, the system for communication within the organisation and the quality of leadership at all levels. Research has shown that financially related motivation systems, such as the payment of safety bonuses, do not necessarily change attitudes, people reverting to normal behaviour when the motivator is removed.

Limitations in human capacity to perceive, attend to, remember, process and act upon information are all relevant in the context of human error. Typical human errors are associated with lapses of attention, mistaken actions, mis-perceptions, mistaken priorities and, in a limited number of cases, wilfulness.

Establishing a safety culture – the principles involved

The main principles involved when establishing a safety culture are generally accepted to be:

(a) the acceptance of responsibility at and from the top, exercised through a clear chain of command, seen to be actual and felt through the organisation;

(b) a conviction that high standards are achievable through proper management;

(c) setting and monitoring of relevant objectives/targets, based upon satisfactory internal information systems;

(d) systematic identification and assessment of hazards and the devising and exercise of preventive systems which are subject to audit and review; in such approaches, particular attention is given to the investigation of error;

(e) immediate rectification of deficiencies; and

(f) promotion and reward of enthusiasm and good results.

(The above list is taken from: Rimington, J R (1989) *The Onshore Safety Regime*, HSE Director General's submission to the Piper Alpha Inquiry.)

Developing a safety culture – essential features

These can be summarised through the guidelines outlined in the CBI's *Developing a Safety Culture*, 1991 (see below).

Several features can be identified from the study which are essential to a sound safety culture. A company wishing to improve its performance will need to judge its existing practices against them.

1. Leadership and commitment from the top which is genuine and visible. This is the most important feature.
2. Acceptance that it is a long-term strategy which requires sustained effort and interest.
3. A policy statement of high expectations and conveying a sense of optimism about what is possible supported by adequate codes of practice and safety standards.
4. Health and safety should be treated as other corporate aims, and properly resourced.
5. It must be a line management responsibility.
6. 'Ownership' of health and safety must permeate all levels of the workforce. This requires employee involvement, training and communication.
7. Realistic and achievable targets should be set and performance measured against them.
8. Incidents should be thoroughly investigated.
9. Consistency of behaviour against agreed standards should be achieved by auditing and good safety behaviour should be a condition of employment.
10. Deficiencies revealed by an investigation or audit should be remedied promptly.
11. Management must receive adequate and up-to-date information to be able to assess performance.

Communications

Much of the answer lies in effective communication on health and safety. Many staff see, for instance, health and safety training as dull, boring, uninteresting or unrelated to their specific tasks. In certain cases they do not understand the reason why certain precautions are enforced by management and safety practitioners. Furthermore, many safety practitioners lack training in communication, seeing the running of health and safety training operations as a chore. In some cases, they may not have the time due to demands of their full-time job.

As with so many areas of performance, there must be communication both vertically and laterally. The Board must set the standards which are both meaningful and measurable. They must communicate these standards down throughout the organisation and ensure such feedback as to enable them to measure and compare performance.

Barriers to communication

A number of barriers can arise at the various phases of the process, in particular:

1. **Barriers to reception**
   Reception of communication can be influenced by:
   (a) the needs, anxieties and expectations of the receiver/listener;
   (b) the attitudes and values of the receiver; and
   (c) environmental stimuli, eg noise.

2. **Barriers to understanding**
   Understanding is a complex process and is affected by:
   (a) the use of inappropriate language, technical jargon;
   (b) the extent to which the listener can concentrate on receiving the data completely, ie variations in listening skills;
   (c) prejudgements made by the listener;
   (d) the ability of the listener to consider factors which may be disturbing or contrary to his ideas and opinions, ie the degree of open-mindedness that he possesses;
   (e) the length of the communication; and
   (f) the degree of knowledge possessed by the listener.

3. **Barriers to acceptance**
   Acceptance of a communication is affected by:
   (a) the attitudes and values of the listener;
   (b) individual prejudices held by listeners;
   (c) status clashes between the sender and the receiver; and
   (d) interpersonal emotional conflicts.

Health and safety training

An important feature of the jobs of health and safety specialists, supervisors and line managers is that of preparing and undertaking short training sessions for their staff on general and specific health and safety issues.

The following points need consideration if such activities are to be successful and effective at conveying the appropriate messages to staff.

1. A list of topics to be covered, eg safe systems of work, manual handling procedures, etc should be developed, and a specific programme should be formulated.
2. Sessions should last no longer than 30 minutes.
3. Visual aids – films, videos, slides, flip charts, etc – should be used extensively.
4. Topics should, as far as possible, be of direct relevance to the group.
5. Participation should be encouraged with an emphasis on identifying possible misunderstandings or concerns that people may have. This is particularly important when introducing a new safety system or operating procedure.
6. Topics should be presented in a relatively simple fashion, using terms that operators can understand. The use of unfamiliar

technical, legal or scientific terminology should be avoided, unless an explanation of such terms is incorporated in the session.

7. Consideration must be given to eliminating any boredom, loss of interest or adverse response on the part of the participants. Talks should be given on as friendly a basis as possible and in a relatively informal atmosphere. Many people respond adversely to a formal classroom situation commonly encountered in staff training activities.

Summary

1. Although most UK health and safety legislation places the duty of compliance firmly on the employer or body corporate, this duty can be discharged by the effective actions of its managers.

2. Studies by the HSE's Accident Prevention Advisory Unit indicate that the vast majority of fatal accidents at work could have been prevented by positive management action.

3. Insufficient attention is paid by managers to the human factors element of safety, in particular those areas of influence on people at work – the organisation, the job and personal factors.

4. Managers need to take into account the organisational characteristics which influence safety-related behaviour.

5. Systems of work and the design of jobs should take into account human capabilities and fallibilities.

6. People are different in terms of how they perceive risk, in their attitude to work, and in what motivates them to work safely.

7. Effective communication and training on health and safety issues is vital in order to ensure good levels of health and safety performance.

References

CBI (1991) *Developing a Safety Culture* CBI, London

Health and Safety Executive (1981) *Managing safety: a review of the role of management in occupational health and safety* HMSO, London

Health and Safety Executive (1985) *Deadly maintenance: A study of fatal accidents at work* HMSO, London.

Health and Safety Executive (1985) *Monitoring safety: An outline report on occupational safety and health by the Accident Prevention Advisory Unit of the Health and Safety Executive* HMSO, London

Health and Safety Executive (1989) *Human factors in industrial safety: Guidance Note HS(G)48* HMSO, London

Rimington, J R (1989) *The Onshore Safety Regime* HSE Director General's submission to the Piper Alpha Inquiry

Stranks, J (1995) *Human Factors and Safety* Pitman, London

Chapter 6
Ergonomics

There are many definitions of the term 'ergonomics'. These include:

(a) the scientific study of the interrelationships between people and their work;
(b) fitting the task to the individual;
(c) the scientific study of work; and
(d) the study of the man–machine interface.

Considerable attention to the principles of ergonomics, ergonomic design, interface design and anthropometry can have significant benefits in reducing stress in the workforce, thereby promoting greater efficiency and reduced manufacturing losses.

The application of ergonomics in working situations

Fundamentally, ergonomics covers four principal areas: the human system, the working environment, the man-machine interface and the total working system. Each of these areas is briefly described below.

1. The human system
This area is concerned with people, in particular the physical aspects, such as stamina, strength and body dimensions, and the psychological aspects of human behaviour, such as perception, learning and reaction to given situations. These features are significant in a wide range of jobs.

2. The working environment
Stress in the working environment can have serious effects on worker performance. Typical environmental stressors are extremes of temperature, inadequate or badly designed lighting, inadequate ventilation, high levels of humidity, noise, vibration, dust, fumes and radiation. The provision of a safe working environment, in addition to being a legal requirement under the HASAWA is a prerequisite for sound levels of worker performance.

3. The man–machine interface
Machine manufacturers and designers frequently produce machinery which places the operator under considerable stress, either through the

design and location of controls or badly designed displays. Good standards of design of this man–machine interface take into account the design of controls, displays, the effects of automation and communication systems, particularly on very large machines, with a view to reducing operator error.

4. Total working system

Factors such as the potential for fatigue and stress are considered at this stage, along with aspects such as work rate and productivity. Specific health and safety features are also considered, in particular the effects of operator error.

These factors may be summarised as in Table 1.

Human characteristics	Environmental factors
Bodily dimensions	Temperature
Strength	Humidity
Stamina	Lighting
Learning	Noise
Mental and physical limitations	Vibration
Perception	Dust, fumes, etc
Reaction	Ventilation
Man–machine interface	Total working system
Controls	Work rate
Displays	Posture
Communications	Fatigue
Automation	Stress
	Productivity
	Accident and ill health
	Health and safety

Table 1   The ergonomic approach to the work situation

Principles of ergonomic design

The elimination of operator error is one of the principal objectives of ergonomic design. The following aspects are taken into account in the design process.

1. Vision

The operator should be able to set and read with ease controls, specific instruments and displays of instruments. This prevents or reduces the fatigue that is so frequently the cause of faulty perception and accidents.

2. Posture

Abnormal working postures increase the potential for fatigue, accidents and long-term injury. All work processes and systems of work should be designed to permit a comfortable posture which reduces excessive strain, for example in the case of VDU operators and lorry drivers. The siting of controls and displays on a wide range of machinery and plant, vehicles and assembly plant is, therefore, crucial to productivity and the prevention of accidents.

3. Layout

In any working area there should be free movement between operating positions, safe access and egress, and unhindered oral and visual communication. Badly planned, congested and over-populated working areas result in operator fatigue, inattention on the part of the operator and increased accident potential in that area.

4. Comfort

Environmental factors have a direct effect on operator comfort, in particular the levels of lighting and ventilation and the percentage relative humidity of the atmosphere (70 per cent is the optimum relative humidity in most cases).

5. Work rate

Ideally, work rates should be set to suit the operator, but need constant reassessment and revision. Movements which are too slow or too fast cause fatigue. This is particularly applicable where operators are engaged in assembly work using a moving track or conveyor.

Controls and displays

Controls regulate a particular machine function to achieve maximum efficiency. Controls can take many forms, eg the hand brake on a vehicle, the stop-start button of a horizontal lathe. Displays, on the other hand, supply information to the operator, eg a VDU screen, petrol gauge in a vehicle. Interface design is concerned with the design and location of controls and displays with the principal objective of minimising or eliminating the potential for operator error. This can be particularly significant in the case of aircraft, large machines, power stations and other complicated machinery and plant where operator error could result in disastrous failure involving loss of life. A number of aspects are important in the design of the man-machine interface.

1. Separation: Displays should be separated from physical controls with, preferably, no operational relationship between them.

2. Order of use: Controls and displays should be arranged in their order of use, eg left to right for start up and right to left for stopping a machine or closing down the plant.

3. Comfort: Where it is not possible to separate controls from displays, they should be mixed to produce a system which can be operated with ease.

4. Function: In the case of large consoles, eg in power stations, controls can be divided according to function. Operator training is essential in this case and is aimed at eliminating the chance of error.

5. Priority: Here the controls most frequently used are sited in key positions. This form of design is appropriate where there is no competition for space.

6. Operator fatigue: Design should take into account the risk of operator fatigue and the likelihood of mistakes being made. Controls should, therefore, be conveniently located with a view to reducing, as far as possible, both visual and postural fatigue.

Visual display units

The two principal problems associated with VDU operation are visual fatigue and postural fatigue. The need to reduce, as far as possible, these forms of fatigue should preferably be considered when establishing the layout and operation of the VDU workstation.

1. Visual fatigue
This can be associated with:

    (a) poor definition of the characters against the background field;
    (b) unsuitable background lighting;
    (c) glare;
    (d) poor legibility, due to factors in the VDU such as 'flicker', 'shimmer' and 'jitter', which are directly related to the refresh rate of the system;
    (e) poor quality source material; and
    (f) visual defects on the part of the operator.

Items (a) to (d) can be corrected through good VDU and workplace design. Very few people, however, have perfect vision, the ability to see varying with age and the presence or absence of visual deficiencies such as myopia (short-sightedness) and hypermetropia (long–sightedness).

Vision screening is recommended, therefore, as a standard feature of a pre-employment health screen for VDU operators.

People who wear spectacles designed for a narrow range of reading distances and those with bifocal or other multifocal lenses may experience some difficulty with tasks involving varying distances. They may find, for instance, that they need to adopt uncomfortable working postures in order to read documents or displayed text satisfactorily. Such individuals may need modifications to their prescription lenses in order to be able to do this type of work, and should consult an optician before undertaking VDU work and whenever discomfort or eye strain is experienced afterwards.

In certain cases, for instance where there is little choice in the positioning of the screen, the use of screen filters may be beneficial. Filters may give a sharper, clearer image, and reduce the eye strain associated with glare and reflection from lights and windows. The use of tinted spectacles, unless prescribed by an optician, is not recommended.

2. Postural fatigue
This is associated with many occupations where staff are frequently sitting for long periods. Postural fatigue can be experienced by way of neck, shoulder and back ache, together with persistent headaches. The incorporation of the recommendations in 'Workstation design' below should alleviate the problem of postural fatigue.

3. Operational stress
The extent of stress experienced by operators will vary, but the following practices are recommended:

(a) rotation of VDU operators in order to take them away from the screen for limited periods; and

(b) trainees should have a short break from VDU operation every one to two hours during their first three months of training.

It should be appreciated that a good logical screen layout can be helpful in the alleviation of operational stress.

4. Work–related upper limb disorders
Keyboard operators may be at risk from suffering upper limb disorders, formerly referred to as repetitive strain injury (RSI). Upper limb disorders cover a range of conditions which affect the soft tissues of the hand, wrist, arm and shoulder. Typical symptoms include pain, restriction of joint movement and swelling of the soft tissues. There is evidence to show that the condition goes through a number of stages, the symptoms intensifying at each stage, thus:

Stage 1 Aching and fatigue in the affected limb in working hours which settles at nights and at weekends.

Stage 2 Recurrent aching and fatigue shortly after work commences and continuing after work ceases.

Stage 3 Persistent aching and fatigue and weaknesses while at rest, plus pain with even non-repetitive movements.

In most cases, there is a need to pursue ergonomic solutions, eg improving workstation layout, which may include job rotation. However, this may not be possible and agreeing to specific screen breaks, where the operator actually moves away from the screen, may be the only solution.

Health and Safety (Display Screen Equipment) Regulations 1992

Poor design and layout of the workstation and lack of attention to environmental factors, eg background lighting, are the principal causes of occupational stress, ie visual and postural fatigue, discomfort and, for some people, annoyance. In certain cases, this lack of attention can lead to one or more work-related upper limb disorders.

The Health and Safety (Display Screen Equipment) Regulations 1992 apply to users and operators of this equipment. A 'user' is an employee who *habitually* uses display screen equipment as a *significant* part of his normal work. An 'operator', on the other hand, is a self-employed person who habitually uses display screen equipment as a significant part of his *normal* work. Typical examples of display screen equipment users are word-processing pool workers, secretaries, data input operators, journalists, tele-sales operators and graphic designers. A receptionist in a hotel would not be classified as a 'user'.

The Regulations, which are accompanied by an HSE Guidance Note, place specific duties on employers. Thus, each employer shall:

(a) perform a suitable and sufficient analysis of workstations used by users and operators; (Regulation 2)

(b) ensure that new workstations meet the requirements laid down in the Schedule to the Regulations; (Regulation 3)

(c) plan the activities of users so that their daily work on display screen equipment is periodically interrupted by breaks or changes of work activity to reduce their workload at that equipment; (Regulation 4)

(d) provide eye and eyesight tests for users, which must be carried out by a competent person; (Regulation 5)

(e) provide adequate health and safety training for users in the use of any workstation upon which he may be required to work; (Regulation 6) and

(f) provide adequate information to ensure that operators and users

Figure 8   VDU workstation principles
(*Source* Health and Safety (Display Screen Equipment) Regulations 1992)

know about all aspects of health and safety relating to their workstations, and about such measures taken by the employer in compliance with his duties under Regulations 2 and 3 as relate to them and their work.

A 'workstation' is defined as an assembly comprising

(a) display screen equipment (whether provided with software determining the interface between the equipment and its operator or user, a keyboard or any other input device);
(b) any optional accessories to the display screen equipment;
(c) any disk drive, telephone, modem, printer, document holder, work chair, work desk, work surface or other item peripheral to the display screen equipment; and
(d) the immediate environment around the display screen equipment.

The analysis of workstations should be based on the criteria outlined in the Schedule below.

The Schedule

This Schedule sets out the minimum requirements for workstations which are contained in the annex to Council Directive 90/270/EEC on the minimum safety and health requirements for work with display screen equipment.

1. Extent to which employers must ensure that workstations meet the requirements laid down in this schedule.

An employer shall ensure that a workstation meets the requirements laid down in this schedule to the extent that:

(a) those requirements relate to a component which is present in the workstation concerned;

(b) those requirements have effect with a view to securing the health, safety and welfare of persons at work; and

(c) the inherent characteristics of a given task do not make compliance with those requirements inappropriate as respects the workstation concerned.

2. Equipment

General comment

The use as such of the equipment must not be a source of risk for operators or users.

Display screen

The characters on the screen shall be well defined and clearly formed, of adequate size and with adequate spacing between the characters and lines.

The image on the screen should be stable, with no flickering or other forms of instability.

The brightness and contrast between the characters and the background shall be easily adjustable by the user, and also easily adjustable to ambient conditions.

The screen must swivel and tilt easily and freely to suit the needs of the operator or user.

It shall be possible to use a separate base for the screen or an adjustable table.

The screen shall be free of reflective glare and reflections liable to cause discomfort to the user.

Keyboard

The keyboard shall be tiltable and separate from the screen, so as to allow the operator or user to find a comfortable working position, avoiding fatigue in the arms or hands.

The space in front of the keyboard shall be sufficient to provide support for the hands and arms of the operator or user.

The keyboard shall have a matt surface to avoid reflective glare.

The arrangement of the keyboard and the characteristics of the keys shall be designed to facilitate the use of the keyboard.

The symbols on the keys shall be adequately contrasted and legible from the design working position.

Work desk or work surface

The work desk or work surface shall have a sufficiently large, low-reflectance surface and allow a flexible arrangement of the screen, keyboard, documents and related equipment.

The document holder shall be stable and adjustable and shall be positioned so as to minimise the need for uncomfortable head and eye movements.

There shall be adequate space for operators or users to find a comfortable position.

Work chair

The work chair shall be stable and allow the user easy freedom of movement and a comfortable position.

The seat shall be adjustable in height.

The seat back shall be adjustable in both height and tilt.

A footrest shall be made available to any user who wishes one.

3. Environment

Space requirements

The workstation shall be designed to provide sufficient space for the user to change position and vary movements.

Lighting

Any room lighting or task lighting provided shall ensure satisfactory conditions and an appropriate contrast between the screen and the background environment, taking into account the type of work and the vision requirements of the operator or user.

Possible disturbing glare and reflections on the screen or other equipment shall be prevented by co-ordinating workplace and workstation layout with the positioning and technical characteristics of the artificial light sources.

Reflections and glare

Workstations shall be so designed that sources of light, such as windows and other openings, transparent or translucid walls, and brightly

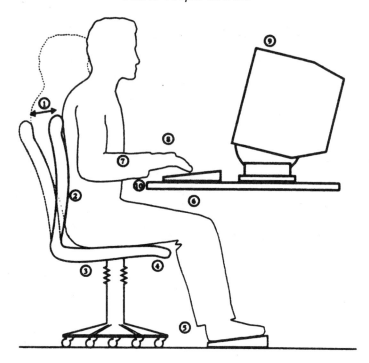

Figure 9 Correct seating and posture
(*Source* Health and Safety (Display Screen Equipment) Regulations 1992)

coloured fixtures on walls cause no direct glare or distracting reflections on the screen.

Windows shall be fitted with a suitable system of adjustable covering to attenuate the daylight that falls on the workstation.

Noise
Noise emitted by equipment belonging to any workstation shall be taken into account when a workstation is being equipped, with a view in particular to minimising distraction.

Heat
Equipment belonging to any workstation shall not produce excess heat which could cause discomfort to operators or users.

Radiation
All radiation with the exception of the visible part of the electromagnetic spectrum shall be reduced to negligible levels from the point of view of the protection of operators' or users' health and safety.

Humidity
An adequate level of humidity shall be established and maintained.

4. Interface between computer and operator/user

In designing, selecting, commissioning and modifying software, and in designing tasks using display screen equipment, the employer shall take into account the following principles:

(a) software must be suitable for the task;
(b) software must be easy to use and, where appropriate, adaptable to the user's level of knowledge or experience; no quantitative or qualitative checking facility may be used without the knowledge of the operators or users;
(c) systems must provide feedback to operators or users on the performance of those systems;
(d) systems must display information in a format and at a pace adapted to operators or users;
(e) the principles of software ergonomics must be applied, in particular to human data processing.

Manual handling

Over 25 million man days are lost annually due to back pain. A substantial proportion of this cost to the country is associated with bad manual handling practices. Disorders of the back include:

(a) strains and tears of tendons, ligaments and muscles;
(b) prolapsed intervertebral disc ('slipped disc');
(c) lumbago, a term descriptive of severe lumbar pain;
(d) rheumatism, particularly in joints;
(e) fracture/compression of the spinal cord; and
(f) hernias.

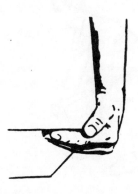

Figure 10   Correct handling – proper grip

Principles of correct handling

1. Correct grip
The correct grip makes use of the palm of the hand and roots of the fingers and thumb. (See Figure 10.) Do not grip with the fingertips as this will lead to strained fingers and muscles in the forearm.

2. Straight back
In order to pick up a load with a straight back one must approach the task by flexing the hips, knees and ankles and the load must be held close to the body. The lift is brought about by the powerful muscles of the legs and not the back, which is kept straight throughout the movement. (See Figure 11.)
    Remember! A bent back is a weak back and can lead to a strained back.

3. Head up
Practise raising the top of the head and this will help you maintain a straight back – an essential movement that should be carried out prior to every lift. This will also enable you to see where you are going.

4. Correct foot position
Always have the feet apart but no wider than the hips, and one foot should be in advance of the other. This leading foot should point in the direction you intend to move. (See Figure 12.)

5. Arms close to the body
Lifting, carrying or pushing with the arms away from the sides of the body results in needless strain being put on the chest, upper back and shoulder muscles. Keep the arms as close to the body as possible.

Figure 11   Correct handling – straight back

Figure 12 Correct handling – proper foot positions

6. Use your body weight
Properly employed, your body weight can be used in moving a load by acting as a counterbalance and thus reducing the amount of muscular effort. (See Figure 13.)

Note: These basic principles should be applied in all manual handling operations. They do, however, need regular practice and training of operators in the techniques.

Figure 13 Correct handling – use your bodyweight

Manual Handling Operations Regulations 1992
These Regulations implement the European 'Heavy Loads' Directive. They supplement the general duties placed upon employers and others by the HASAWA and the broad requirements of the MHSWR 1992, and are supported by Guidance issued by the HSE.
    The Regulations define a number of terms as follows:

Injury – does not include injury caused by any toxic or corrosive substance which:

(a) has leaked or spilled from a load;
(b) is present on the surface of a load but has not leaked or spilled from it; or
(c) is a constituent part of a load.

Load – includes any person and any animal.

Manual handling operations – means any transporting or supporting of a load (including the lifting, putting down, pushing, pulling, carrying or moving thereof) by hand or by bodily force.

The following duties of employers are laid down under Regulation 4:

1. Each employer shall
    (a) so far as is reasonably practicable, avoid the need for his employees to undertake any manual handling operations at work which involve a risk of their being injured;
    (b) where it is not reasonably practicable to avoid the need for his employees to undertake any manual handling operations at work which involve a risk of their being injured, he must:
        (i)   make a suitable and sufficient assessment of all such manual handling operations to be undertaken by them, having regard to the factors which are specified in the Schedule;
        (ii) take appropriate steps to reduce the risk of injury to those employees arising out of their undertaking any such manual handling operations to the lowest level reasonably practicable;
        (iii) take appropriate steps to provide any of those employees who are undertaking such manual handling operations with general indications and, where it is reasonably practicable to do so, precise information on:
            (aa) the weight of each load, and
            (bb) the heaviest side of any load whose centre of gravity is not positioned centrally.
2. Any assessment such as referred to in para 1(b)(i) of this Regulation shall be reviewed by the employer who made it if:
    (a) there is reason to suspect it is no longer valid; or
    (b) there has been a significant change in the manual handling operations to which it relates.

If changes to an assessment are required, as a result of any such review, the relevant employer shall make them.

The Regulations further require that each employee, while at work, shall make full and proper use of any system of work provided for his use by his employer in compliance with Regulation 4(1)(b)(ii) of these Regulations.

Guideline weights for lifting
The following diagram shows recommended guideline weights for lifting.

Schedule 1
Factors to which the employer must have regard and questions he must consider when making an assessment of manual handling operations.

1. The tasks
Do they involve:
- holding or manipulating loads at distance from trunk?
- unsatisfactory bodily movement or posture, especially:

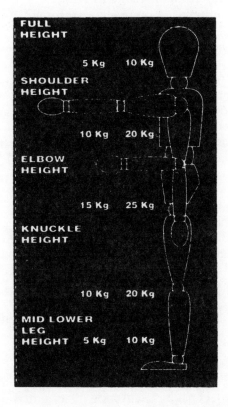

Figure 14 Diagram shows guideline weights for lifting
(*Source*: HSE)

- twisting the trunk?
- stooping?
- reaching upwards?
- excessive movement of loads, especially:
  - excessive lifting or lowering distances?
  - excessive carrying distances?
- excessive pushing or pulling of loads?
- risk of sudden movement of loads?
- frequent or prolonged physical effort?
- insufficient rest or recovery periods?
- a rate of work imposed by a process?

2. The loads

Are they:

- heavy?
- bulky or unwieldy?
- difficult to grasp?
- unstable, or with contents likely to shift?
- sharp, hot or otherwise potentially damaging?

3. The working environment

Are there:

- space constraints preventing good posture?
- uneven, slippery or unstable floors?
- variations in level of floors or work surfaces?
- extremes of temperature or humidity?
- conditions causing ventilation problems or gusts of wind?
- poor lighting conditions?

4. Individual capability

Does the job:

- require unusual strength, height, etc?
- create a hazard to those who might reasonably be considered to be pregnant or have a health problem?
- require special information or training for its safe performance?

5. Other factors

Is movement or posture hindered by personal protective equipment or clothing?

Management action

Management should consider an Action Plan aimed at eliminating or reducing accidents and back pain associated with poor manual handling practices. Such an Action Plan should incorporate the following:

1. Identification of all manual handling operations and the risks to operators.
2. Examination of current sickness absence records associated with manual handling, eg days lost through back strain, fatigue, handling injuries, and identification of the costs to the organisation.
3. Examination of the nature of the loads handled by people in terms of size, shape, weight, rigidity, etc.
4. Consideration of the availability of existing manual handling equipment and the need for improved equipment in the future.
5. Take into account the physical and mental characteristics of future staff in terms of their ability to handle loads as a feature of pre-employment health screening.
6. Undertake training of staff, both at the induction stage and on a regular basis, and highlight the risk of back injury through the use of posters, films and other forms of increasing awareness.

Summary

1. Ergonomics is concerned with people and the work that they do, the potential for error in their work and the effects of work on their health.

2. Good standards of ergonomic design are essential in order to reduce operator error.

3. Design of the 'man–machine interface' should consider, particularly, the use and layout of controls and displays.

4. The introduction of the 'new technology' in offices has created a number of problems in terms of operator visual fatigue and postural fatigue. VDU workstation design should consider these aspects with a view to their prevention.

5. Over 25 million working days are lost through back pain. Back pain is the second most common reason, next to the common cold, for people being absent from work. The need for well-established manual handling procedures is, therefore, paramount if this loss to the country is to be reduced.

References

Bell, C R (1974) *Men at Work* George Allen & Unwin Ltd, London

Health and Safety Executive (1992) *Health and Safety (Display Screen Equipment) Regulations 1992 and Guidance on Regulations* HMSO, London

Health and Safety Executive (1992) *Manual Handling Operations Regulations 1992 and Guidance on Regulations* HMSO, London

Health and Safety Executive (1992) *Lighten the load: Guidance for employees on musculoskeletal disorders* HSE Information Centre, Sheffield

Health and Safety Executive (1993) *Working with VDUs* HSE Information Centre, Sheffield

Health and Safety Executive (1993) *Getting to grips with manual handling* HSE Information Centre, Sheffield

Health and Safety Executive (1993) *Ergonomics at Work* HSE Information Centre, Sheffield

International Labour Organisation (1984) *ILO Encyclopaedia: Lifting and Carrying* ILO, Geneva

McCormick, E J (1976) *Human Factors Engineering* McGraw-Hill, New York

Shackel, B (1974) *Applied Ergonomics Handbook* IPC Science and Technology Press, Guildford

Singleton, W T (1974) *Man-Machine Systems* Penguin Books, Harmondsworth

Singleton, W T (1976) *Human Aspects of Safety* Keith Shipton Developments, London

Chapter 7
Stress at Work

The amount of time lost from work in the United Kingdom is typically in excess of 300 million days per annum. At least 10 per cent of these days lost are attributed to what is officially referred to as 'psychoneurosis'. To this figure of 30 million working days lost through psychoneurosis must be added further time lost through 'psychosomatic' complaints, namely physical illnesses that originated in, or have been exacerbated by, psychological and stress-related problems, and all uncertified absence. Such losses represent a significant cost to all forms of organisation, and do not include the cost of low productivity and decreased efficiency due to low motivation, increases in alcohol and drug consumption, and time lost through 'presenteeism' ie being physically present at work, but mentally absent.

What is stress?
Stress is a term which is rarely clearly understood. Various definitions have been put forward over the years, as follows:

1. Any influence that disturbs the natural equilibrium of the living body.
2. The common response to attack (Hans Selye, 1936).
3. A feeling of sustained anxiety which, over a period of time, leads to disease.
4. A psychological response which follows failure to cope with problems.

Generally, a stressful circumstance is one with which an individual is unable to cope successfully, or believes he cannot cope successfully, and which results in unwanted physical, mental or emotional responses. Stress implies some form of demand on the individual, it can be perceived as a threat, it can produce the classic 'flight or fight' response, it may create physiological imbalance and can certainly affect individual performance. It is particularly concerned with how people cope with changes in their lives at work, at home and in other circumstances. It should be appreciated, however, that not all stress is bad. We all need a certain amount of stress (positive stress) in order to cope with life situations on an on-going basis.

Classification of stressors
A stressor produces stress. There are many forms of stressor, namely:

1. Physical stressors – extremes of temperature, lighting, ventilation and humidity, noise and vibration.
2. Chemical stressors – dangerous chemicals: gases, vapours, dusts, etc
3. Biological stressors – bacteria, viruses, etc.

However, most people associate stress with social or psychological stress which may be brought about, perhaps, by isolation, rejection, pressure and a general overloading of the body systems (distress). The demands on people at work vary substantially. Some demands may be related to the actual work that they do or the factors surrounding that work, including:

(a) psychological demands – machine-paced work, the quality of supervision, hazards, monotony of the task;
(b) physical demands – the effort required, as in manual handling activities, the potential for fatigue and exposure to hazardous substances;
(c) demands related to the construction of displays and controls on machinery – display screen equipment, fork-lift trucks, machinery;
(d) environmental demands – noise, pollution, poor lighting, etc;
(e) working hours – shift work, unsocial hours, night work, the frequency of breaks; and
(f) payment arrangements – piece work systems, compliance with quality standards.

Sources of stress among managers may be associated with many factors – their role in the organisation, career development, the organisational structure and climate, relationships within the organisation and certain factors which are specific to the job. There may also be demands from outside the organisation. Cooper and Marshall (1978) demonstrated the conflict between the demands of the organisation and external demands, eg those of the family, as a significant cause of management stress (see diagram on page 105). The concept of the organisational boundary, with the manager astride that boundary, is well established. Personal factors are important in this case in terms of individual personality, a person's tolerance for ambiguity, ability to cope with change, level of motivation and specific behavioural patterns.

Role theory

Role theory views most large organisations as comprising systems of interlocking roles. These roles relate to what people do and what others expect of them. Problems arise as a result of:

1. **Role ambiguity** This is the situation where the role holder has insufficient information for the adequate performance of his role, or

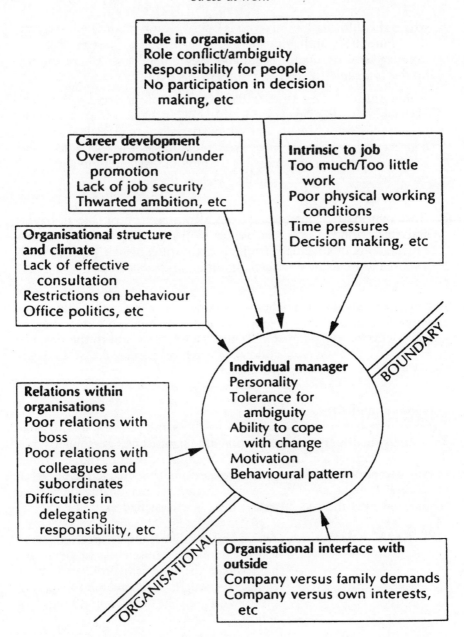

Figure 15   Sources of managerial stress
(*Source*: Cooper and Marshall, 1978)

where the information received is open to more than one interpretation. Potentially ambiguous situations are in posts where there is a time lag between action taken and visible results, or where the role holder is unable to see the results of his actions.

2. **Role conflict**   This arises where members of the organisation, who exchange information with the role holder, have different expectations of his role. (Health and safety specialists frequently suffer this conflict situation.) Each may exert pressure on the role holder and, commonly, satisfying one expectation could make compliance with the other expectations difficult. This is the classic 'servant of two masters' situation.

3. **Role overload**   This results from a combination of role ambiguity and role conflict. The role holder works harder to clarify normal expectations or to satisfy conflicting priorities which are frequently impossible to achieve within the time limits specified.

Research has shown that when experience of role conflict, ambiguity and overload is high, then job satisfaction is low. This may well be coupled with worry and anxiety. These factors may add to the onset of stress-related diseases and conditions such as peptic ulcers, coronary heart disease and nervous breakdowns.

Personality and stress

No two people necessarily respond to the same stressor in the same way. Individual personality factors are significant. Personality was defined by Allport as; 'the dynamic organisation within the individual of the psychophysical systems that determine his characteristic behaviour and thought'. Various types and traits of personality have been established over the last 30 years, these are classified as follows:

1. **Type 'A' – Ambitious**   Active and energetic; impatient if he has to wait in a queue; conscientious; maintains high standards; time is a problem – there is never enough; frequently intolerant of those who may be slower in thought or action.

2. **Type 'B' – Placid**   Quiet; very little worries them; put their worries into things they can alter or control and leave others to worry about the rest.

3. **Type 'C' – Worrying**   Nervous; highly strung; not very confident of self-ability; anxious about the future and of being able to cope.

4. Type 'D' – Carefree Loves variety; often athletic and daring; very little worries them; not concerned about the future.

5. Type 'E' – Suspicious Dedicated and serious; very concerned with other people's opinions of them; do not take criticism kindly and tend to dwell on such criticism for a long time; distrust most people.

6. Type 'F' – Dependent Bored with their own company; sensitive to surroundings; rely on others a great deal; people who interest them are oddly unreliable; they find that the people they really need are boring; do not respond easily to change.

7. Type 'G' – Fussy Punctilious; conscientious and like a set routine; do not like change; any new problem throws them because there are no rules to follow; conventional and predictable; great believers in authority.

Research indicates that most people combine traits of more than one of these 'types', and so the above definitions can only be used as a guide. The type most at risk to stress is Type A.

Women at work
Women can be subject to many stressors at work which are not suffered by their male counterparts. Whilst sexual harassment is a common cause of stress amongst women, other causes of stress include:

(a) performance-related pressures;
(b) lower rates of pay;
(c) the problem of maintaining dependants at home;
(d) lack of encouragement from superiors, including not being taken seriously;
(e) discrimination in terms of advancement;
(f) sex discrimination and prejudice;
(g) pressure from dependants at home;
(h) career-related dilemmas, including whether to start a family or whether to marry or live with someone;
(i) lack of social support from colleagues;
(j) being single and labelled as an oddity; and
(k) lack of domestic support at home.

Management should be aware of the various forms of stress women are exposed to whilst at work. Wherever necessary measures should be taken to reduce stress. Special attention should be given to cases of sexual harassment, which can be handled using disciplinary action.

The effects of stress

Stress effects vary considerably from person to person. Typical effects of stress are headaches, insomnia, fatigue, overeating, constipation, nervousness, minor accidents, palpitations, indigestion and irritability. Many more effects and symptoms could be added to this list.

The two principal psychological effects of stress are anxiety and depression.

1. **Anxiety**   This is a state of tension coupled with apprehension, worry, guilt, insecurity and a constant need for reassurance. It is accompanied by psychosomatic symptoms, such as profuse perspiration, difficulty in breathing, gastric disturbances, rapid heart beat, frequent urination, muscle tension or high blood pressure. Insomnia is a reliable indicator of a state of anxiety.

2. **Depression**   This has been defined as 'a sadness which has lost its relationship to the logical progression of events' (David Viscott, *American Psychiatrist*). Its milder form may be a direct result of a crisis in work relationships. Severe forms may exhibit biochemical disturbances, and the extreme form can lead to suicide. Another definition is 'a mood, characterised by feelings of dejection and gloom, and other permutations, such as feelings of hopelessness, futility and guilt'.

Typical stressful conditions

1. Too heavy or too light a workload.
2. A job which is too difficult or too easy.
3. Working excessive hours, eg 60 or more hours per week.
4. Conflicting job demands – the 'servant of two masters' situation.
5. Too much or too little responsibility.
6. Poor human relationships.
7. Incompetent superiors, in terms of their ability to make decisions, their level of performance and their job knowledge.
8. Lack of participation in decision-making and other activities where a joint approach would be beneficial.
9. Middle-age vulnerability associated with reduced career prospects or the need to change career, the threat of redundancy or premature retirement.
10. Over-promotion or under-promotion.
11. Interaction between work and family commitments.
12. Deficiencies in interpersonal skills.

Coping strategies

There are a number of ways that people can commonly deal with the emotional and physical aspects of stress at work. These are generally through relaxation training, physical exercise and, in certain cases, the use of drugs. Relaxation training may take the form of progressive muscular relaxation, brief relaxation exercises and several forms of meditation eg mantra meditation.

From an organisational viewpoint, stress at all levels can have a serious effect on performance. Potentially stressful organisations are those:

(a) which are large and bureaucratic;
(b) in which there are formally prescribed rules and regulations;
(c) where there is conflict between positions and people;
(d) where people are expected to work hard for long hours;
(e) where no praise is given;
(f) where the general culture is classed as 'unfriendly'; and
(g) where conflict can arise between normal work and external interests, eg the family.

There is a need, therefore, for organisations to do certain things if they are to reduce stress in the workplace. A Stress Management Action Plan should incorporate the following:

1. Recognition of the causes and symptoms of stress at all levels.
2. Decisions on the need to do something about it.
3. Identification of the group or groups who may be affected by stress at work.
4. Examination and evaluation by interview or questionnaire to determine the causes of stress.
5. Analysis of problem areas.
6. Decision on appropriate strategies, eg training, time management, counselling of and support for individuals, revision of management policies in certain cases.
7. Implementation of a Stress Management Programme taking the above factors into account.

Conclusion

This chapter has endeavoured to give a broad overview of the problem of stress at work, a subject which is rarely considered. The most difficult task is getting people to recognise the existence of the stress response within individuals and within the organisation and that their decisions could be stressful for other people. Many managers still adopt the Victorian maxim 'If you can't stand the heat, get out of the kitchen!'

Such a response is totally unhelpful to people going through stressful events in their lives, be they associated with the work situation or private life.

Summary

1. Stress is a common feature of most people's lives and the causes of stress are many and varied. It is most commonly associated with changes in people's lives, some of which may be brought about by the organisation.

2. There is a need within organisations for a greater understanding of the stress response and the causes of stress.

3. Stress reduction strategies should be considered at boardroom level and implemented wherever necessary.

4. The costs of stress-related ill health can be substantial in terms of time lost for conditions diagnosed as 'anxiety state', 'depression' and 'nervous breakdown'.

References

Allport, G W (1961) *Pattern and Growth in Personality* Holt, New York

Bond, M and Kilty, J (1982) *Practical Methods of Dealing with Stress* University of Surrey, Guildford

Cooper, C L, Cooper, R D and Eaker, L H (1988) *Living with Stress* Penguin Books, London

Cox, T (1978) *Stress* Macmillan Press, London

Health and Safety Executive (1993) *Mental Distress at Work*, HSE Information Centre, Sheffield

Orlans, V and Shipley, P (1983) *A Survey of Stress Management and Prevention Facilities in a Sample of UK Organisations* Birkbeck College, University of London

Selye, H (1936) *The Stress of Life*, rev. ed. 1976, McGraw Hill, New York

Conclusion to Part 2

Human factors are becoming increasingly a significant feature of health and safety management. The need to recognise the various influences on people at work, in particular those characteristics of an organisation which influence safety-related behaviour, cannot be over-emphasised. In the design of jobs and safe systems of work consideration must be given to the potential for human error, human capabilities and fallibilities and the causes and effects of stress.

We do not plan our working systems in many cases, particularly with regard to the effective use of human resources. The application of ergonomic principles, including safe manual practices, would bring about great reductions in the stress associated with many tasks. Companies who adopt ergonomic principles in their working systems have been able to show reductions in accidents and ill health with the resulting improvements in productivity and profitability.

Part 3
Occupational Health

Chapter 8
Occupational Diseases and Conditions

Occupational health, as an area of occupational medicine, is concerned fundamentally with two aspects, namely:

(a) the relationship of work to health; and
(b) the effects of work upon the worker.

Poor standards of occupational health practice by companies frequently result in employees contracting occupational diseases and conditions, such as dermatitis, noise-induced hearing loss (occupational deafness), pneumoconiosis or one of the occupational cancers. One of the frequent criticisms levelled at companies is that, while their occupational safety procedures are well established, systems for the prevention of occupational ill health are either non-existent or, alternatively, limited to certain routine medical examinations, such as pre-employment medical examinations, examinations prior to entry into the company pension scheme or, specifically, on an annual basis, restricted to executives and senior managers.

Occupational diseases

These is a group of diseases and conditions contracted as a result of a particular employment. As such, they may be 'prescribed' under the Social Security Act 1975, in which case benefit is payable to those suffering from them, and 'reportable' under RIDDOR.

Prescribed diseases
Prescribed diseases are listed in Schedule 1 of the Social Security (Industrial Injuries) (Prescribed Diseases) Regulations 1985. Each prescribed disease is related to a specific occupation. A disease is 'prescribed' if:

(a) it ought to be treated, having regard to its causes and incidence and other relevant considerations, as a risk of occupation and not as a risk common to all persons; and
(b) it is such that, in the absence of special circumstances, the attribution of particular cases to the nature of the employment can be established with reasonable certainty.

(Social Security Act 1975, Section 76(2))

Reportable diseases

These are diseases which are reportable to the enforcement agency, eg HSE, local authority, under RIDDOR (Regulation 5). In this case, a wide range of diseases is listed and qualified by a particular work activity, together with a common description of the disease. Typical examples include poisoning by carbon disulphide, methyl bromide, etc; skin diseases, such as folliculitis; occupational asthmas; pneumoconiosis; and certain infections, such as hepatitis. Such diseases must be reported to the enforcement agency on Form 2508A.

Principal causes and effects

RIDDOR classifies occupational diseases on the basis of causation and the work activity. The Social Security (Industrial Injuries) (Prescribed Diseases) Regulations classify in a similar way on the basis of occupation.

Occupational diseases and conditions can be classified in the following ways.

Conditions due to physical agents
1. Temperature – heat stroke, heat cataracts.
2. Lighting – miner's nystagmus.
3. Radiation – radiation sickness, arc eye, burns.
4. Noise – noise-induced hearing loss.
5. Vibration – vibration-induced white finger.
6. Pressure – decompression sickness.
7. Dust – pneumoconiosis, including silicosis, coal worker's pneumoconiosis, occupational asthma, occupational cancers.
8. Repetitive movements – writer's cramp.
9. Manual work – beat elbow, beat knee, beat hand.

Conditions due to biological agents
1. Contact with infected animals – anthrax, brucellosis, glanders.
2. Contact with blood or blood products – viral hepatitis.
3. Vegetable-borne infections – farmer's lung (aspergillosis).
4. Contact with rodents – leptospirosis.

Conditions due to chemical agents
1. Poisonings – the use or handling of, or exposure to, the fumes, dust or vapour of a wide range of chemical substances, including lead, manganese, phosphorus, arsenic and mercury.
2. Occupational cancers – associated with exposure to tar, pitch, bitumen, mineral oils, aromatic amines, such as alpha-naphthylamine and beta-naphthylamine, and vinyl chloride monomer (VCM).
3. Specific damage to organs and systems – the use or handling of, or exposure to the fumes of, or vapours containing solvents, such as carbon tetrachloride, trichloromethane (chloroform) and chloromethane (methyl chloride).

Conditions due to work activities

1. Job movements – writer's cramp, 'beat' conditions eg beat hand, beat elbow, beat knee, traumatic inflammation of the tendons or associated tendon sheaths of the hand or forearm ie tenosynovitis, work-related upper limb disorders (repetitive strain injury).
2. Friction and pressure – bursitis, cellulitis.

The Employment Medical Advisory Service (EMAS)

The EMAS is fundamentally the Medical Services Division of the HSE. It comprises a national network of around 150 doctors and nurses accountable to a number of Senior EMAs and the Director of Medical Services. Approximately 22,000 statutory medical examinations are undertaken every year, together with 90,000 such examinations carried out by company occupational physicians who are specified as 'appointed doctors' under the system.

Under these statutory requirements:

(a) the employer is officially notified of the fitness of the worker to undertake employment;

(b) the employee has a duty to undergo examinations;

(c) the employer cannot lawfully continue to employ any worker who is found to be unfit;

(d) the employee must be removed from that particular job for the prescribed period and transferred to alternative work if it is available; and

(e) the outcome of the medical examination must be recorded in a specific health register by the employer.

Sickness absence control

The cost of sickness absence to all forms of commercial undertaking can be substantial if procedures are not established and followed. Over 200 million man days are lost each year due to absence from work, a large proportion of which is attributed to sickness. With approximately 20 million people employed in the United Kingdom, this amounts to between 10 and 11 days' absence per person per year – ie an absence rate of 4.6 per cent. Absence can be associated with frequent periods of short-term sickness and prolonged periods of sickness absence. Failure to produce sick notes and the taking of frequent short unconnected periods of absence can, in certain cases, result in an employee being dismissed. Much will depend upon that employee's terms of contract, the statement of terms and conditions, features of the company sick pay and absence scheme and the specific action taken by the employee. Non-compliance with the company scheme may result in breach of contract with subsequent withholding of his pay. The company, on investigating the reasons for frequent periods of short-term sickness absence, may decide the employee is guilty of misconduct and, on the basis of a national agreement, take disciplinary action. The disciplinary action could be a verbal or written warning, or alternatively, dismissal.

Long-term sickness absence hinges around 'capability' to work, rather than misconduct. Capability to work is assessed by reference to skill, aptitude, health or any other physical or mental quality. Where the employer makes a decision to terminate employment on the basis of incapability to work, he should first consider the potential for his being taken to a tribunal on the grounds of unfair dismissal. The following questions should, therefore, be asked:

1. Were offers of alternative employment considered?
2. Was the employee consulted sufficiently about his problems, future employment prospects and the possibility of dismissal?
3. Was the latest period of sickness absence prior to receiving the warning of the possibility of dismissal? Was a further warning appropriate in this case?
4. Has there been a recent opportunity for the employee to comment on his health problem and ability to work? Was an independent medical opinion sought?
5. Has the employer sought medical advice over this situation?
6. Has the employer investigated all relevant matters prior to the dismissal decision?
7. How long, apart from the period of illness, would the employment be likely to last?
8. Can the employer wait any longer for the employee to return to work? (A balance must be achieved between the position of the employee, the interests of the company and the need to be fair.)
9. How important is it for this employee to be replaced?
10. Has the employee been consulted on the final step in the procedure?

Once these questions have been answered satisfactorily, a decision may need to be taken as to whether or not employment should be terminated. The four questions below should then be considered:

1. Were the terms of the contract of employment, including sick pay provisions, fulfilled?
2. What is the nature of the employment, with particular reference to a key position?
3. What is the nature of the illness/injury? How long has it continued and what are the prospects of recovery?
4. What is the total period of employment?

At completion of this exercise, the employee must be interviewed, informed of management's decision and the result confirmed in writing.

In the event of an appeal to an Employment Appeals Tribunal, the tribunal would be seeking evidence of the following having been carried out by the employer:

1. A fair review of the employee's attendance record and the reasons for periods of absence.
2. Evidence of the employee having received appropriate warnings after he had been given an opportunity to make representations.

Where there has been no improvement in the attendance record, it is likely that, in most cases, the employer would be justified in treating the persistent absences as sufficient reason for dismissing the employee.

It will be seen that sickness absence control hinges around a clearly established procedure linked to the allocation of sick pay.

Sickness absence procedure

A typical company sickness reporting and pay procedure should take the form shown below.

IN ORDER TO QUALIFY FOR SICK PAY ALL STAFF MUST

1. Notify their supervisor/foreman by 10.00 am (or at least two hours prior to the commencement of a shift) on the first day of absence, giving an indication as to the expected length of absence.

2. Where sick between one and three days, employees must complete the company self-certification form; if not submitted, payment can be withheld.

3. Where sick between four to seven days, employees must submit DSS self-certification form; if not submitted, payment can be withheld.

4. If sick for more than seven days a doctor's certificate must be submitted; if not submitted, payment can be withheld.

5. On return to work, the employee must notify his supervisor of his intention to return by 2.00 pm on the day prior to return, or for night- shift workers, by 12 noon of the day of the shift.

Note:

1. In cases 2 and 3 above, submission of the doctor's certificate should be made on the day of return to work.

2. A doctor's certificate should be submitted within at least 10 days of the commencement of the period of absence. If the certificate is not dated from the first day of absence, the employee must complete the self-certification form.

Sickness absence monitoring

All supervisors/foremen should maintain a day-to-day record of employees' attendance at work. This can be done on a monthly basis whereby, on each working day, a record is made of those employees who are present, those who are absent for a particular reason (eg annual leave, rest day, etc), and those who are absent through sickness or unexplained reasons. On this basis it is relatively simple to identify patterns of short-term sickness absence among the workforce and to instigate early investigatory processes to determine the causes of such absences. Employees should be reminded regularly of the need to follow the sickness absence procedure above, and disciplinary action should be taken against persistent defaulters.

The procedure should ensure that employers act as follows:

1. They tackle the problem face to face.
2. They observe and record the problem.
3. They find out the facts.
4. They agree a plan of action.

The ultimate objectives of a sickness absence control system are to:

(a) recognise and measure absenteeism associated with ill health;
(b) identify causes of such absence in the workplace;
(c) deal fairly with employees who are frequently absent due to ill health;
(d) set up procedures to discourage absenteeism; and
(e) recognise when the company needs specialist help in dealing with the problem.

Health surveillance

The on-going monitoring of people's health is an important feature of a number of industries eg the mining, lead, food and chemical industries. It may take the form of pre-employment health examinations and examinations at specified periods by an occupational health nurse, specific medical examinations by an occupational physician, or specific forms of biological monitoring, as in the case of people who could be exposed to organo-phosphorus compounds.

Generally, health surveillance concentrates on two principal groups of workers:

(a) those actually or potentially at risk by virtue of the type of work they undertake during their employment eg radiation workers; and
(b) those at risk of developing further ill health or disability by virtue

of their present state of health eg people who have been exposed to excessive levels of noise.

Under the COSHH Regulations health surveillance may be identified as a necessary protective procedure following the completion of a health risk assessment for a particular hazardous substance used in the workplace.

Policies on smoking at work

With the increasing attention that has been given to the risks associated with passive smoking, organisations should consider the development of policies on smoking at work.

The problem has been identified mainly in poorly ventilated open-plan offices and among employees who may suffer from some form of respiratory complaint, eg asthma or bronchitis. Other people may complain of soreness of the eyes, headaches and stuffiness.

In 1988 the HSE's booklet 'Passive Smoking at Work' drew attention to the concept of passive smoking, whereby non-smokers inhale environmental tobacco smoke from burning cigarettes, cigars and pipes exhaled by smokers. Work carried out by the Independent Scientific Committee on Smoking and Health identified a small but measurable increase in risk from lung cancer for passive smokers of between 10 per cent and 30 per cent. The booklet also suggests that passive smoking could be the cause of one to three extra cases of lung cancer every year for each 100,000 non-smokers who are exposed throughout life to other people's smoke.

A number of issues are involved. The cost to industry of people smoking at work has never been measured in terms of time lost through smoking-related diseases and ill health, fires caused through careless smoking and, in many cases, time wasted in indulging in the practice. On the other hand, for some people, smoking may be considered a minor form of addiction. Imposing a total ban on smoking, without consultation or assistance to give up the habit, could impose severe stress on these people. So how could companies seek to regulate this problem with a view to eventually phasing out smoking at work?

The first step is the development of a Policy on Smoking at Work. This should state the intention of the company to eliminate smoking in the workplace by a specific date, the legal requirements on the employer to provide a healthy working environment, and the fact that smoking is bad for the health of smokers and non-smokers alike. The 'organisation and arrangements' for implementing the policy should state the individual responsibilities of managers in the various stages of the operation, as well as the responsibilities of the staff in complying with the policy.

On publication of the policy statement, the operation should proceed in clear stages, commencing with a questionnaire to all managers seeking information on: the number of employees on site; the number of smokers; the number who have given up in the last year; possible problems anticipated through operation of the eventual ban; local efforts made to assist smokers to give up smoking; facilities available to smokers, for example designated smoking areas; and costs incurred in helping people to give up smoking, eg counselling, hypnotherapy. Such collated information will give an indication of the current state of play on smoking in the organisation.

The second stage is concerned with the provision of training, information, instruction, health education, therapy, propaganda and counselling on the health risks associated with smoking, with a view to raising the awareness of all concerned. This should be backed up by the marking of 'No Smoking' areas and the restriction of smoking to exempted rooms and areas. Job applicants should also be advised of the policy at the interview stage.

Some managers may be concerned about the risk of industrial action against the policy and of the problem of staff who persistently breach same. In certain extreme cases, dismissal may be the only solution. A recent industrial tribunal decision supports this view. In Rogers *v* Wicks & Wilson, an employee, who subsequently resigned and claimed unfair dismissal when his employer announced a forthcoming ban on smoking, was held not to have been unfairly dismissed on the basis that employees do not have a contractual right to smoke. Moreover, where such bans are introduced with sufficient warning and consultation with staff, an employer cannot be said to have acted unreasonably.

Summary

1. Poor standards of health control in the workplace can result in employees contracting occupational disease and conditions.

2. The differences between 'prescribed' and 'reportable' diseases should be recognised and understood.

3. The majority of occupational diseases are associated with physical, chemical and biological agents present at work.

4. Staff of the EMAS have powers to take action against employers and occupiers of premises in certain cases where there may be a risk to the health of workers, including the compulsory medical examination of workers.

5. An organisation's sickness absence costs can be substantial if no procedure for monitoring and regulating them is operated.

References

Atherley, G R C (1978) *Occupational Health and Safety Concepts* Applied Science Publishers Ltd, London

Department of Health and Social Security (1980) *Social Security (Industrial Industries) (Prescribed Diseases) Regulations 1985 (SI 1985 No. 967) and subsequent Amendment Regulations 1986 & 1987* HMSO, London

Department of Health and Social Security (1983) *Notes on the Diagnosis of Occupational Diseases* HMSO, London

Health and Safety Executive (1981) *Health Surveillance by Routine Procedures, Guidance Note MS 18* HMSO, London

Health and Safety Executive (1988) *Passive Smoking at Work*, HSE Information Centre, Sheffield.

Reporting of Injuries, Diseases and Dangerous Occurrences Regulations 1985 (SI 1985 No 2023) HMSO, London

Stranks, J (1995) *Occupational Health and Hygiene* Pitman, London

Chapter 9
First Aid

The provision of facilities for first aid treatment in workplaces is a basic legal requirement under the FA, OSRPA and other principal protective legislation.

Legal considerations apart, however, there is clearly a case for as many people as possible being trained, irrespective of their functions, in basic first aid procedures. Such knowledge could be instrumental in saving the life of a person whatever the circumstances. This philosophy has been promoted by many organisations, including the national first aid organisations, following recent disasters involving both workers and members of the public.

Current legal requirements are covered by the Health and Safety (First Aid) Regulations 1981, made under the HASAWA, an ACOP and guidance issued by the HSE.

Health and Safety (First Aid) Regulations 1981

Under these Regulations, 'first aid' is defined as meaning:

(a) in cases where a person will need help from a medical practitioner or nurse, such treatment necessary to preserve life and minimise the consequences of injury and illness until such help is obtained; and
(b) treatment of minor injuries which would otherwise receive no treatment or which do not need treatment by a medical practitioner or nurse.

There are three general duties under the Regulations, namely:

(a) the duty of the employer to provide first aid arrangements;
(b) the duty of the employer to inform his employees of these arrangements; and
(c) the duty of the self-employed person to provide first aid equipment.

These duties are expanded further in the ACOP and Guidance Notes issued with the Regulations. There are four specific criteria which the employer must take into account in deciding what are adequate and appropriate arrangements for first aid in the workplace, namely:

(a) the number of employees;
(b) the nature of the undertaking;
(c) the size of the establishment, and distribution of employees; and
(d) the location of the establishment and of the employees' places of work.

An adequate number of suitable persons able to give first aid must be provided, and these persons must be adequately trained and hold an appropriate first aid qualification approved by the Health and Safety Executive. In certain circumstances, eg high-risk activities, additional training may be necessary.

Note: It should be stressed, however, that cover must be adequate at all times when people are at work and there is clearly a case, therefore, for having more trained first-aiders available than shown above in order to cover for holidays, sickness, shift working, etc.

In situations where it is not necessary to appoint a first-aider, an appointed person must be designated who can take charge of situations, ie call a doctor and/or ambulance, in the event of serious injury or major illness.

Many employees work away from a main location. In these cases the employer has the responsibility to make adequate and appropriate first aid facilities available for these staff and ensure that this takes place. Furthermore, where employees work alone or in small groups, where the work involves travelling long distances, or where employees may be using potentially dangerous equipment or machinery, small travelling first aid kits must be provided.

Contents of first aid boxes and kits are specified in the Regulations. They must be checked and replenished as necessary.

First aid rooms

The code of practice provides that an employer should generally provide a suitably equipped and staffed first aid room only where 400 or more employees are at work. However, in the case of:

(a) establishments with special hazards;
(b) construction sites with more than 250 persons at work; and
(c) when access to casualty centres, or emergency facilities, is diffi-cult, eg owing to distance or inadequacy of transport facilities;

a first aid room should be provided.

The role of first-aiders

The ACOP places special emphasis on the need for employers to avail themselves of first-aiders trained in specific techniques required by their undertaking. In certain cases, for instance where there may be a risk of cyanide poisoning or there may be a need for oxygen for resuscitation purposes, training should be undertaken by organisations approved by the HSE.

The provision of trained first-aiders, furthermore, is related to specific circumstances, as in the case of shift work, where there must be adequate coverage for each shift. In the case of low risk establishments eg offices, shops, there need normally be no first-aider where fewer than 150 employees are at work, with at least one first-aider for 150 employees or more. For establishments with greater risk, eg factories, farms, the employer should provide one first-aider per 50–150 employees, and an additional first-aider where there are more than 150 employees. Where there are fewer than 50 employees, an employer must in any case provide an appointed person.

First aid equipment

Where first aid boxes form part of an establishment's first aid provision, they should contain only those items which a first aider has been trained to use. Sufficient quantities of each item should always be available in every first aid box or container. In most cases, these will be:

(a) one guidance card;
(b) twenty individually wrapped sterile adhesive dressings (assorted sizes) appropriate to the work environment (which may be detectable for the food and catering industries);
(c) two sterile eye pads with attachment;
(d) six individually wrapped triangular bandages;
(e) six safety pins;
(f) six medium sized individually wrapped sterile unmedicated wound dressings (approx. 10 cm $\times$ 8 cm);
(g) two large sterile individually wrapped unmedicated wound dressings (approx. 13 cm $\times$ 9 cm); and
(h) three large sterile individually wrapped unmedicated wound dressings (approx. 28 cm $\times$ 17.5 cm).

Where mains tap water is not available for eye irrigation, sterile water or sterile normal saline (0.9%) in sealed disposable containers should be provided. Each container should hold at least 300 ml and should not be reused once the sterile seal is broken. At least 900 ml should be provided. Eye baths/eye cups/refillable containers should not be used for eye irrigation.

Where an employee has received additional training in the treatment of specific hazards which require the use of special antidotes or special equipment, these may be stored near the hazard area or may be kept in the first aid box.

Travelling first aid kits

In the case of employees who regularly work away from their employer's establishment in isolated locations, or where they are involved in travelling long distances in remote areas from which access to accident and emergency facilities may be difficult, it may be necessary for first aid equipment to be carried by, or made available to, employees where potentially dangerous tools or machinery are used. The equipment should be suitable for the numbers involved and the potential hazards to which employees are exposed.

The contents of travelling first aid kits should be appropriate for the circumstances in which they are used. At least the following should be included:

(a) card giving the general first aid guidance;
(b) six individually wrapped sterile adhesive dressings;
(c) one large sterile unmedicated dressing;
(d) two triangular bandages;
(e) two safety pins; and
(f) individually wrapped moist cleaning wipes.

Guidance on first aid

The Regulations require the following information to be kept in every first aid box, usually in the form of a card.

HEALTH AND SAFETY (FIRST AID) REGULATIONS 1981

General first aid guidance for first aid boxes

NOTE: TAKE CARE NOT TO BECOME A CASUALTY YOURSELF WHILE ADMINISTERING FIRST AID. BE SURE TO USE PROTECTIVE CLOTHING AND EQUIPMENT WHERE NECESSARY. IF YOU ARE NOT A TRAINED FIRST-AIDER, SEND IMMEDIATELY FOR THE NEAREST FIRST-AIDER WHERE ONE IS AVAILABLE.

Advice on treatment

If the assistance of medical or nursing personnel will be required, send for a doctor or nurse (where they are employed at the workplace) or ambulance immediately. When an ambulance is called, arrangements should be made for it to be directed to the scene without delay.

Priorities

1. Breathing
If the casualty has stopped breathing, resuscitation must be started at once BEFORE ANY OTHER TREATMENT IS GIVEN and should be continued until breathing is restored or until medical, nursing or ambulance personnel take over.

Mouth-to-mouth resuscitation

2. Bleeding

If bleeding is more than minimal, control it by direct pressure – apply a pad of sterilised dressing or, if necessary, direct pressure with fingers or thumb on the bleeding point. Raising a limb if the bleeding is sited there will help reduce the flow of blood (unless the limb is fractured).

3. Unconsciousness

Where the patient is unconscious, care must be taken to keep the airway open. This may be done by clearing the mouth and ensuring that the tongue does not block the back of the throat. Where possible, the casualty should be placed in the recovery position.

Recovery position

4. Broken bones

Unless the casualty is in a position which exposes him to further danger, do not attempt to move a patient with suspected broken bones or injured joints until the injured parts have been supported. Secure so that the injured parts cannot move.

5. Other injuries

    (a) Burns and scalds: Small burns and scalds should be treated by flushing the affected area with plenty of clean cool water before applying a sterilised dressing or clean towel. Where the burn is large or deep, simply apply a dry sterile dressing. (*NB* Do not burst blisters or remove clothing sticking to the burns or scalds.)

    (b) Chemical burns: Remove any contaminated clothing which shows no sign of sticking to the skin and flush all affected parts of the body with plenty of clean cool water ensuring that all the

chemical is so diluted as to be rendered harmless. Apply a sterilised dressing to exposed damaged skin and clean towels to damaged areas where the clothing cannot be removed. (*NB* Take care when treating the casualty to avoid contamination.)

(c) Foreign bodies in the eye: If the object cannot be removed readily with a clean piece of moist material, wash with clean cool water. People with eye injuries which are more than minimal must be sent to hospital with the eye covered with a sterilised eye pad.

(d) Chemical in the eye: Flush open the eye AT ONCE with clean cool water: continue for at least 5 to 10 minutes and, in any case of doubt, even longer. If the contamination is more than minimal, send the casualty to hospital.

(e) Electric shock: Ensure that the current is switched off. If this is impossible, free the person, using heavy duty insulating gloves (to BS 697: 1977) where these are provided for this purpose near the first aid container, or using something made of rubber, dry cloth or wood, or a folded newspaper; use the casualty's own clothing if dry. BE CAREFUL not to touch the casualty's skin before the current is switched off. If breathing is failing or has stopped, start resuscitation and continue until breathing is restored or medical, nursing or ambulance personnel take over.

(f) Gassing: Move the casualty to fresh air BUT MAKE SURE THAT WHOEVER DOES THIS IS WEARING SUITABLE RESPIRATORY PROTECTION. If breathing has stopped, start resuscitation and continue until breathing is restored or until medical, nursing or ambulance personnel take over. If the casualty needs to go to hospital make sure a note of the gas involved is sent with him.

General
(a) Hygiene
When possible, wash your hands before treating wounds, burns or eye injuries. Take care in any event not to contaminate the surfaces of dressings.
(b) Treatment position
Casualties should be seated or lying down while being treated.

(c) Record keeping
An entry must be made in the Accident Book, eg Form BI 510, of each case.

(d) Minor injuries
Casualties with minor injuries of a sort they would attend to themselves if at home, may wash their hands and apply a small sterilised dressing from the container.

(e) First aid materials
Each article used from the container should be replaced as soon as possible.

Principles of first aid

The aims of first aid treatment are to sustain life, to prevent deterioration in an existing condition, and to promote recovery.

Principal aspects of such treatment are resuscitation (restoration of breathing), control of bleeding and the prevention of collapse. Reference to the 'general guidance' which must be shown in first aid boxes, clearly reinforces these principal aspects.

All staff, irrespective of whether they are designated first-aiders, should be familiar with the resuscitation procedure shown on page 131. The procedures, in card form, are available from RoSPA (see Figure 16).

The role of first-aiders

A first-aider is a person who has received training and who holds a current first aid certificate from an organisation or employer whose training and qualifications for first-aiders are approved by the HSE.

The definition of 'first aid' in the Regulations, or the alternative definition below, are a useful guide in deciding on the first aid arrangements necessary for a particular enterprise. First aid is:

'the skilled application of accepted principles of treatment on the occurrence of an accident or in the case of sudden illness, using facilities and materials available at the time'.

The identification of first aid needs, in terms of the number of, and degree of training for first-aiders, provision of facilities and the arrangements for ensuring 100 per cent first aid provision at all times, is a legal requirement. Employers must identify such needs and review them on a regular basis.

Summary

1. The Health and Safety (First Aid) Regulations 1981, together with the ACOP and Guidance Notes, provide guidance for employers on the arrangements necessary for first aid in the workplace.

2. Specific provision must be made for high-risk locations, people who work away from the main location and the giving of information to staff.

3. All staff should be aware of the principles of first aid and basic first aid procedures.

4. The skilled application of first aid procedures, such as restoration of

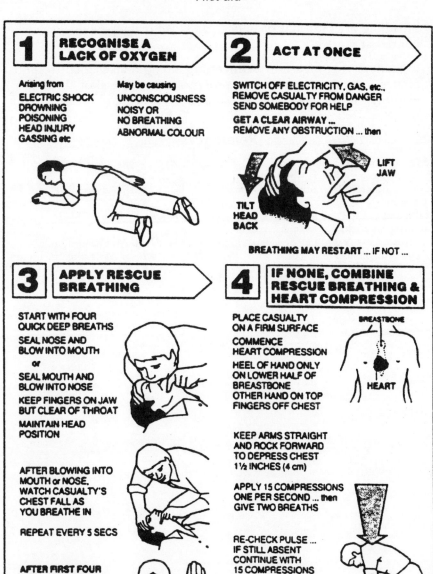

Figure 16 Resuscitation procedure display card
(*Source* Royal Society for the Prevention of Accidents)

breathing or control of bleeding, has saved many people's lives when injured at work.

5. Procedures for first aid treatment, including management responsibilities for ensuring adequate provision of same, should be clearly identified in the company Statement of Health and Safety Policy.

References

Health and Safety Commission (1990) *Approved Code of Practice: Health and Safety (First Aid) Regulations 1981* HMSO, London

Health and Safety Executive (1981) *First Aid at Work, HSS Booklet HS(R)11* HMSO, London

Health and Safety Executive (1992) *First Aid Needs in your Workplace*, HSE Information Centre, Sheffield

Health and Safety (First Aid) Regulations 1981, (SI 1981 No 917) HMSO, London

St John Ambulance Association and Brigade (1972) *First Aid Manual* SJAB, London

Chapter 10
Dangerous Substances

Dangerous substances are used in a wide range of industrial and commercial activities and, over the years, many workers have contracted occupational diseases through exposure to such substances. Typical examples include occupational dermatitis (non-infective dermatitis), chemical poisonings (eg by derivatives of arsenic, phosphorus and lead), occupational cancers and the group of diseases known as the pneumoconioses (eg asbestosis, siderosis, coal worker's pneumoconiosis and silicosis). The potential for contracting an occupational disease will vary according to the potential for harm of the substance concerned, its form (eg solid, liquid, gas, dust etc), the route of entry into the body, the precautions taken (both physically by the employer and personally by the worker), the dose received on a short-term, medium-term and long- term basis, and the degree of susceptibility to the substance of the individual.

Other substances, while not directly entering the body, can be dangerous – eg flammable, explosive, cryogenic substances, contact with which can result in burns, physical damage such as major injuries, and skin irritation.

Chemicals (Hazard Information and Packaging for Supply) (CHIP) Regulations 1994

These Regulations implement a series of European Community Directives on the classification, packaging and labelling of dangerous substances and preparations. They came into force on 31 January 1995. The objective of the CHIP Regulations is to help protect people and the environment from the ill-effects of chemicals. Under the Regulations, suppliers are required to:

(a) identify the hazards (or dangers) of the chemicals they supply;
(b) give information about the hazards to the people who are supplied; and
(c) package the chemicals safely.

A *supplier* is someone who supplies chemicals as part of a transaction, eg manufacturers, importers and distributors.

Supply requirements

The supply requirements implement a series of directives adopted by the Member States of the European Community (EC) on classification,

133

packaging and labelling of dangerous substances and preparations. (*Substances* are pure chemicals, like ethanol or water, and *preparations* are mixtures of chemicals, like paints or gin. The term *chemical* covers substances and preparations.)

Scope of the CHIP Regulations

This is one of the largest packages of Regulations derived from international agreements. It is relevant to all chemicals, from the most complicated and esoteric to household commodities like paint or bleach. The CHIP Regulations, developed to educate companies to take the right precautions, concern anyone who supplies or consigns dangerous chemicals in Great Britain or offshore. This includes not only large companies but also small manufacturers who concentrate on making one or more common chemicals or who retail chemicals to the public.

The CHIP Regulations are also the foundation for other health and safety and environmental provisions. For example, substances classified under the CHIP Regulations as being harmful to health fall within the scope of the COSHH Regulations. The safety data sheets required by the CHIP Regulations are a major source of information for those who have to prepare COSHH assessments.

Certain substances identified as dangerous by the CHIP Regulations fall within the scope of the Control of Major Industrial Accident Hazards (CIMAH) Regulations 1984.

The package of legislation and supporting guidance for the CHIP Regulations consists of:

(a) the Regulations
(b) the Approved Supply List: Information Approved for the Classification, Packaging and Labelling of Substances and Preparations Dangerous for Supply;
(c) the Approved Guide to the Classification and Labelling of Substances and Preparations Dangerous for Supply; and
(d) the Approved Code of Practice on Safety Data Sheets for Substances and Preparations Dangerous for Supply.

Classification requirements

The fundamental requirement is for a supplier to decide if the chemical is dangerous or not. If it is dangerous, then it must be classified, and it is an offence to supply a chemical before this classification is complete.

The first step in classification is to put the chemical into a category of danger. There are three main categories of dangerous chemicals, which are further sub-divided:

(a) substances and preparations dangerous because of their physical or chemical properties:
   (i)   explosive;
   (ii) oxidizing;
   (iii) extremely flammable;
   (iv) highly flammable;
   (v)   flammable;
(b) substances and preparations dangerous because of their health effects:
   (i)   very toxic;
   (ii)   toxic;
   (iii) harmful;
   (iv)   corrosive;
   (v)   irritant;
   (vi)   sensitising;
   (vii) carcinogenic (substances which cause cancer);
   (viii) mutagenic (substances which cause inherited changes);
   (ix)   toxic for reproduction (substances which cause harm to the unborn);
(c) substances dangerous for the environment.

Although suppliers are responsible for classification, they do not have to undertake the classification themselves. However, the supplier must ensure that the classification has been carried out by a competent person.

Some 1,400 of the more common dangerous chemicals have been classified – details are given in the Approved Supply List. Each substance is identified by its chemical name or its unique international number, and the classification and other information can be read directly from the List.

Where a substance is not on the Approved Supply List, then suppliers have to classify the substances themselves on the basis of the properties described in the Regulations. The competent person must find and assemble the available information on the chemical, but it will not be necessary to carry out any new tests. Once the data has been assembled, it can be compared with the criteria in the Approved Guide to the Classification and Labelling of Substances and Preparations Dangerous for Supply. The competent person must then decide whether the substance is dangerous and what classification (if any) is appropriate.

Preparations must be classified in the same categories, except for the classification for environmental effects, which is not required. Since one of the objectives of the CHIP Regulations is to avoid unnecessary animal testing, there is a method for calculating a preparation's category of danger which avoids the practice altogether.

Safety data sheets

One of the important requirements is that safety data sheets be provided for dangerous chemicals which are supplied for work. A safety data sheet must be provided by the supplier to the recipient, and must contain information about the chemical to enable the recipient to take the right precautions. The ACOP: Safety Data Sheets for Substances and Preparations Dangerous for Supply gives advice.

The safety data sheet is invaluable in enabling an employer to carry out a COSHH assessment. However, it is not a substitute for an assessment. Safety data sheets will describe the hazards of the chemicals, but only the user can assess the risk in the workplace.

Safety data sheets must be provided whether the chemical is sold in bulk or in packages. However, they do not have to be provided when chemicals are sold for private use through shops, due to the existence of comparable legislation covering such cases.

Labelling

Packaging containing dangerous chemicals must be properly labelled. The aim is to inform anyone handling the package or using the chemical about its hazards and to advise on the precautions. For workers the label is the supplement to information provided by the employer; for others it is a major way of propagating information.

The label must always contain details about:

(a) the supplier;
(b) the chemical;
(c) the category of danger; and
(d) risk phrases and safety phrases.

Risk phrases and safety phrases are standard phrases set out in the CHIP Regulations. *Risk phrases* describe the dangers of the chemicals in more detail, eg 'May cause cancer' or 'Toxic by inhalation'. *Safety phrases* tell the user what to do or not to do with the chemical, eg 'Keep away from children' or 'Do not empty into drains'.

A *warning symbol* is also required on most labels, eg a skull and crossbones or a picture of an explosion. Suppliers are responsible for using the correct label.

Labelling follows classification. For substances, the first step is to look in the Approved Supply List. Where the substance is not listed, or if the chemical is a preparation, then the label is derived from the classification using the advice in the Approved Guide to the Classification and Labelling of Substances and Preparations Dangerous for Supply.

Packaging

The Regulations require chemicals to be packaged safely to withstand the conditions of supply.

Hazardous substances and preparations – classification

The following classifications are detailed in Schedule 1 of the CHIP Regulations

Category of danger	Property (See Note 1)	Symbol – letter
1. Physico-Chemical Properties		
Explosive	Solid, liquid, pasty or gelatinous substances and preparations which may react exothermically without atmospheric oxygen, thereby quickly evolving gases. Under defined test conditions these can detonate, quickly deflagrate or explode if heated while being partially confined.	E
Oxidising	Substances and preparations which give rise to an exothermic reaction in contact with other substances, particularly those which are flammable.	O
Extremely flammable	Liquid substances and preparations having an extremely low flash point and a low boiling point. Also, gaseous substances and preparations which are flammable in contact with air at ambient temperature and pressure.	F+
Highly flammable	The following substances and preparations: (a) substances and preparations which may become hot and finally catch fire in contact with air at ambient temperature without any application of energy; (b) solid substances and preparations which may readily catch fire after brief contact with a source of ignition and which continue to burn or to be consumed after removal of the source of ignition; (c) liquid substances and preparations having a very low flash point; (d) substances and preparations which, in contact with water or damp air, evolve highly flammable gases in dangerous quantities. (See Note 2)	F
Flammable	Liquid substances and preparations having a low flash point.	None
2. Health Effects		
Very toxic	Substances and preparations which in *very low quantities* can cause death, acute or chronic damage to health when inhaled, swallowed or absorbed via the skin.	T+
Toxic	Substances and preparations which in *low quantities* can cause death, acute or chronic damage to health when inhaled, swallowed or absorbed via the skin.	T

Table 2 cont.

Category of danger	Property (See Note 1)	Symbol – letter
Harmful	Substances and preparations which may cause death, acute or chronic damage to health when inhaled, swallowed or absorbed via the skin.	Xn
Corrosive	Substances and preparations which may, on contact with living tissues, *destroy* them.	C
Irritant	Non-corrosive substances and preparations which through immediate, prolonged or repeated contact with the skin or mucous membrane, may cause *inflammation.*	Xi
Sensitising	Substances and preparations which, if they are inhaled or penetrate the skin, are capable of eliciting a reaction by *hypersensitisation,* so that on further exposure to the substance or preparation, characteristic adverse effects are produced.	
Sensitising by inhalation		Xn
Sensitising by skin contact		Xi
Carcinogenic (*see Note 3)*	Substances and preparations which, if they are inhaled, ingested or penetrate the skin, may induce *cancer* or increase its incidence.	
Category 1		T
Category 2		T
Category 3		Xn
Mutagenic (*See Note 3)*	Substances and preparations which, if they are inhaled, ingested or penetrate the skin, may induce *hereditable genetic defects* or increase their incidence.	
Category 1		T
Category 2		T
Category 3		Xn
Toxic for reproduction (*See Note 3)*	Substances and preparations which, if they are inhaled, ingested or penetrate the skin, may produce or increase the incidence of *non-hereditable adverse effects* in the progeny, and/or an impairment of male or female reproductive functions or capacity.	
Category 1		T
Category 2		T
Category 3		Xn
Dangerous for the environment (*See Note 4)*	Substances which, if they were to enter into the environment, would/could present an immediate or delayed danger for one or more components of the environment.	N

Table 2 Categories of danger

Corrosive

Oxidizing

Explosive

Toxic

Harmful/ Irritant

Highly Flammable

Figure 17   Hazard warning symbols

Notes
1. As further described in the *Approved Classification and Labelling Guide*.
2. Preparations packed in *aerosol dispensers* shall be classified as *flammable* in accordance with the additional criteria set out in Part II of this Schedule.
3. The categories are specified in the *Approved Classification and Labelling Guide*.
4. (a) In certain cases specified in the *approved supply list* and the *Approved Classification and Labelling Guide*, substances known to be *dangerous for the environment*, do not require to be labelled with the symbol for this category of danger.
   (b) This category of danger does not apply to preparations.

Routes of entry of toxic substances into the body

Inhalation
This is the principal route of entry of dangerous substances, in the form of dusts, gases, vapours, mists, etc, into the body, and accounts for approximately 90 per cent of all cases of ill health associated with exposure to toxic substances. The effects may be:

(a) acute or immediate, as in gassing accidents associated with the evolution of carbon monoxide, chlorine and nitrous oxide; or

(b) chronic, as with exposure to chlorinated hydrocarbons, lead compounds and numerous dusts.

Pervasion (Absorption)

This is normally associated with occupational dermatitis, which may be linked with exposure to primary irritants or secondary cutaneous sensitisers. Primary irritants generally cause dermatitis at the site of contact when permitted to act for a sufficient length of time and in sufficient concentration, eg acids, alkalis and solvents. Secondary cutaneous sensitisers, however, do not necessarily cause skin changes on first contact, but bring about a specific sensitisation of the skin. When further contact occurs, dermatitis will develop at the site of the second contact. Nickel, rubber, certain plants and a wide range of chemical substances are known skin sensitisers.

Ingestion

This implies entry through the mouth into the digestive system. Typical entry is through consumption of contaminated food and drink.

Injection and implantation

Injection of dangerous substances directly may take place through a forceful breach of the skin, perhaps as a result of an accident. On the other hand, implantation implies a controlled insertion of a substance into the body.

Aspiration

This is the introduction of a liquid into the lungs, such as organic solvents. This can result in severe inflammation of the lungs (pneumonitis). Moreover, aspiration can occur during vomiting, particularly where a patient may be unconscious or semi-conscious.

Handling and storage of dangerous substances

Handling and storage of dangerous substances can present both general and specific problems, in particular the risk of incompatible reactions associated with poor standards of housekeeping, spillages, inadequate separation of substances, poor labelling (particularly of small quantities which may have been transferred from bulk containers), the need for clearly defined first aid treatments covering a range of dangerous substances, clearly defined dispensing procedures and high standards of supervision and control. Above all, information on stored products which may be potentially dangerous should be readily available and staff should be trained in its interpretation. Such information must be provided in the form of safety data sheets on all substances and preparations used at work by the manufacturers, importers and suppliers.

Safety data sheet/Product information

The CHIP Regulations specify that obligatory information under the following headings must be provided in a safety data sheet:

1. Identification of the substance/preparation
2. Compostion/information on ingredients
3. Hazards identification
4. First aid measures
5. Fire fighting measures
6. Accidental release measures
7. Handling and storage
8. Exposure controls/Personal protection
9. Physical and chemical properties
10. Stability and reactivity
11. Toxicological information
12. Ecological information
13. Disposal considerations
14. Transport information
15. Regulatory information
16. Other information

Control of Substances Hazardous to Health (COSHH) Regulations 1994

Industry uses some 40,000 different substances. Each one could be hazardous and will, therefore, fall within the scope of the COSHH Regulations. The Regulations apply to nearly all companies and organisations, from the major chemical manufacturer to the smaller craft workshops, but the following industries are especially involved:

(a) major manufacturers and bulk users of chemical substances;
(b) users of substances in circumstances most likely to involve high exposure levels, eg spraying activities;
(c) the so-called 'dusty trades', including the ceramics and refractories industries, quarrying, foundries and metal manufacturing/finishing processes; and
(d) users of processes which generate substances hazardous to health in appreciable quantities, eg certain welding, cutting, grinding, milling or sieving operations.

The principles outlined in the COSHH Regulations are supported by a number of ACOPs and Guidance Notes issued by the HSE.

Principal requirements of the COSHH Regulations

1. The employer must take account of the properties of substances used at work, in particular the EEC Inventory of over 5,000 substances and the EEC Directive on hazardous agents.
2. Employers have the following duties:
   - (a) to make health risk assessments;
   - (b) to control exposures;
   - (c) to carry out monitoring; and
   - (d) to arrange for health surveillance, in particular:
      - (i) health examinations;
      - (ii) environmental monitoring; and
      - (iii) the maintenance of control measures, eg exhaust ventilation systems.

Practical implementation of the Regulations implies that the 'responsible person' – ie the employer or controller of premises, manufacturer and/or supplier of substances – must know about the potential for harm to employees or others of any substances used, sold, supplied or disposed of as part of his undertaking.

A 'substance' is defined as 'any natural or artificial substance, whether solid, liquid, gas or vapour, and includes human pathogens'. Lead and asbestos are specifically excluded from the Regulations.

A 'substance hazardous to health' means any substance (including any preparation) which is:

- (a) listed in Part I of the approved supply list as dangerous for supply within the meaning of the CHIP Regulations 1993 and classified as very toxic, toxic, harmful, corrosive or irritant;
- (b) a substance specified in Schedule 1 (which lists substances assigned maximum exposure limits) or for which the HSC has approved an occupational exposure standard; (see HSE Guidance Note EH40 'Occupational Exposure Limits)
- (c) a biological agent;
- (d) dust of any kind, when present in a substantial concentration in air; and
- (e) any other substance arising from work which may be hazardous to health.

Health risk assessments

Perhaps the most significant duty on the employer under the Regulations is the requirement to make a health risk assessment. Such an assessment should identify:

- (a) the risks posed to the health of the workforce;
- (b) the measures necessary to control exposure to those hazards; and

(c) other action that may be necessary to achieve compliance with Regulations 8–12.

Moreover, the assessment must be reviewed forthwith if there is reason to suspect that the assessment is no longer valid, or there has been a significant change in the work to which the assessment relates and, where as a result of the review, changes in the assessment are required, those changes must be made.

The starting point for most assessments is the hazard data information supplied by the manufacturer, supplier or importer, which must be provided in accordance with Section 6 of HASAWA. Whilst there is no standard format for a health risk assessment, the following information should be available at the completion of the exercise:

(a) the risk involved in the use of the substance (toxic, corrosive, harmful or irritant), together with the route(s) of entry of that substance into the body, eg by inhalation of its vapour;
(b) the engineering or other controls necessary to prevent health risks arising;
(c) health and/or medical surveillance procedures necessary;
(d) environmental monitoring procedures necessary to be undertaken in the workplace;
(e) extent of the information, instruction, training and supervision necessary for operators using the substance; and
(f) the record keeping requirements, eg maintenance and test of local exhaust ventilation systems.

Company compliance procedures
Procedures should take place in two stages, thus:

1. Preliminary procedure

(a) Prepare an inventory of all the substances used for production, maintenance, cleaning and laboratory analysis.
(b) Identify the point of use for each material.
(c) Ensure information on each substance is adequate. If the information is inadequate, the manufacturer/supplier should be required to provide more comprehensive information.
(d) Once the list is established, a programme of assessment and control of hazardous substances should be commenced.

2. Final procedure

(a) Identify the risks through health risk assessment procedures.
(b) Assess and evaluate the risks.

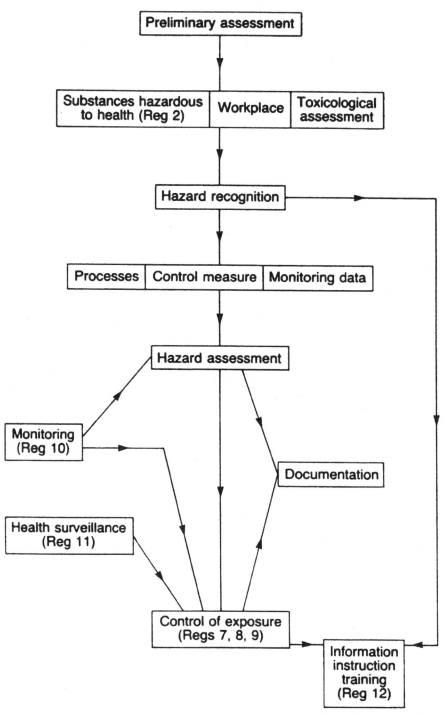

Figure 18   COSHH action plan

(c) Control exposures.
(d) Ensure use of control measures.
(e) Maintain control measures.
(f) Institute workplace monitoring.
(g) Institute health surveillance procedures.
(h) Provide information, instruction and training for all staff on a regular basis, and at the induction stages for new employees.

It should be appreciated by employers that health records relating to individual exposures may have to be kept for 30 years.

Reference should be made to the 'COSHH Action Plan' shown in Figure 18 when considering an approach to compliance with the COSHH Regulations.

Summary

1. A wide range of potentially dangerous substances is used in industry.

2. Reference to the descriptive phrases specified in the CHIP Regulations should be made when considering the relative hazards of substances.

3. The route of entry of a hazardous substance is significant in its potential for harm.

4. Procedures should be established for the handling and storage of dangerous substances, including systems for the acquisition of information, and the training of employees in hazards associated with the use of dangerous substances.

5. The COSHH Regulations 1994 place significant duties upon employers and users of dangerous substances, together with manufacturers and suppliers. Procedures for ensuring compliance with the COSHH Regulations should be established, commencing with an inventory of all chemicals currently in use, and the implementation of continuing monitoring and assessment procedures.

References

Atherley, G R C, (1978) *Occupational Health and Safety Concepts* Applied Science Publishers, London

Chemicals (Hazard Information and Packaging for Supply) Regulations 1994 HMSO, London

Control of Substances Hazardous to Health Regulations 1994 HMSO, London

Health and Safety Commission (1995) *Approved Code of Practice: Control of Carcinogenic Substances* HMSO, London

Health and Safety Commission (1995) *Approved Code of Practice: Control of Substances Hazardous to Health* HMSO, London

Health and Safety Commission (1995) *Approved Code of Practice: Control of Biological Agents* HMSO, London

Health and Safety Commission (1988) *Approved Code of Practice: Control of Substances Hazardous to Health in Fumigation Operations* HMSO, London

Health and Safety Executive (1992) *COSHH Assessments* HMSO, London

Health and Safety Executive (1988) *Hazard and Risk Explained: Control of Substances Hazardous to Health Regulations 1988* (Leaflet IND(G) 67(L)), HMSO, London

Health and Safety Executive (1988) *Introducing Assessment: A Simplified Guide for Employers* (Leaflet IND(G) 64(L)), HMSO, London

Health and Safety Executive (1988) *Substances for Use at Work: The Provision of Information* (Guidance Note HS(G) 26), HMSO, London

Health and Safety Executive (under constant revision) *Environmental Hygiene Guidance Note Series: EH/18 Toxic substances: a precautionary policy; EH/22 Ventilation of the workplace; EH/26 Occupational skin diseases: health and safety precautions; EH/40 Occupational exposure limits; EH/42 Monitoring strategies for toxic substances; EH/44 Dust in the workplace: general principles of protection* HMSO, London

Sax, N I (1979) *Dangerous Properties of Industrial Materials* Reinhold, New York

Stranks, J (1997) *Handbook of Health and Safety Practice* Pitman, London

Conclusion to Part 3

The HASAWA has brought about great improvements in occupational safety. Improvements in the occupational health field have, however, been much slower, and the number of days lost through occupational disease and ill health remains unacceptably high, with substantial costs to companies and the nation as a whole.

There is a need, therefore, for organisations to examine their sickness absence rates, to control sickness absence more effectively, and to use the services of occupational health practitioners in order to reduce the toll of ill health associated with the work that people do.

Most companies use dangerous substances in one form or another, and the uncontrolled use of these substances can result in health risks to staff. The COSHH Regulations have the principal objective of eliminating or controlling these risks in order to protect the health of workers. The establishment and implementation of procedures to comply with these Regulations is essential, including health risk assessments for all identified dangerous substances, health surveillance, environmental monitoring and the development of control measures. Where a company does not have trained professionals to undertake this work, they will have to buy in this expertise from outside. Training of operators in the safe use of dangerous substances is essential.

Part 4
Safety Technology

Chapter 11
Engineering Safety

The relative safety aspects of machinery, plant and equipment have always been one of the more significant areas of health and safety enforcement and practice. The law relating to machinery and equipment is covered by the Provision and Use of Work Equipment Regulations (PUWER) 1992, which replace much of the former legislation, ie the FA, OSRPA and Regulations, made under these Acts. Only two terms are defined in these Regulations:

Work equipment means any machinery, appliance, apparatus or tool, and any assembly of components which, in order to achieve a common end, are arranged and controlled so that they function as a whole.

Use in relation to work equipment means any activity involving that equipment, including starting, stopping, programming, setting, transporting, repairing, modifying, maintaining, servicing and cleaning.

Work equipment, thus, can include a dumper truck, a power press, a ladder, a hammer, an electric drill and a tractor. The definition is extremely broad. PUWER covers a number of general duties and states that every employer:

(a) shall ensure work equipment is suitable; (Regulation 5)
(b) shall ensure that work equipment is maintained in an efficient state, in efficient working order and in good repair; (Regulation 6)
(c) shall pay attention to any specific risks created by the work equipment (Regulation 7)
(d) shall provide users with adequate health and safety information and, where appropriate, written instructions pertaining to the use of the work equipment; (Regulation 8)
(e) shall ensure that all users, managers and supervisors have received adequate health and safety training with particular reference to correct working methods, any risks involved and the precautions necessary in the use of the equipment; (Regulation 9)
(f) shall ensure that specific measures are taken to prevent risks arising from certain identified dangerous parts of machinery; (Regulation 11) and

(g) shall take measures to prevent the exposure of any person to a specified hazard, such as work equipment catching fire or over-heating. (Regulation 12)

Principles of machinery safety

In any assessment of machinery safety, two factors must be considered:

(a) The mechanical factors, in terms of design, operation and relia-bility of guards and safety devices; and
(b) the human factors, with regard to physical operation of the machine (the ergonomic aspects are significant here), systems of work, the potential for operator error, routine maintenance procedures and waste removal.

Strategies directed at eliminating the hazards associated with machinery should take account of the principal causes of injury. For instance, a person may be injured while working with machinery through:

(a) coming into contact with it, or being trapped between the machinery and any material in or at the machinery or any fixed structure;
(b) being struck by, or becoming entangled in or by, any material in motion in the machinery;
(c) being struck by parts of the machinery ejected from it; and
(d) being struck by material ejected from the machinery.

(*Source* BS 5304: 1988 'Safeguarding of machinery')

Machinery safety design features should take into account the following forms of machinery hazard with the principal objective of eliminating such hazards. Assessments of existing machinery should also take these hazards into account.

Traps

Traps in machinery, if unguarded, can result in a wide range of major injury accidents, including hand, arm and finger amputations. Traps take three principal forms:

(a) *In-running nips:* This form of trap is created where a moving belt or chain meets a roller or toothed wheel respectively, and at the point where revolving gears/toothed wheels, rollers or drums meet. Typical examples of in-running nips are shown in Figure 19.
(b) *Reciprocating traps:* These are encountered in the vertical and horizontal motion of certain machines, such as presses.

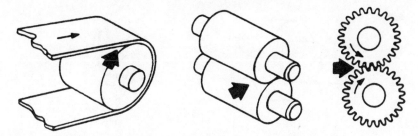

Figure 19   In-running nips: A – Between belt and pulley
B – Between two rollers
C – Between toothed wheels
(Source BS 5304: 1988 'Safeguarding of machinery')

(c) *Shearing traps:* This form of machinery hazard produces a guillotine effect whereby a moving part traverses a fixed part of a
machine or, alternatively, two moving parts traverse each other,
with an action similar to that of garden shears. (See Figure 20.)

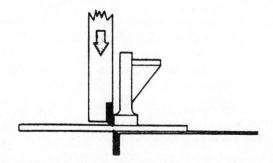

Figure 20   Shearing traps – A moving part traversing a fixed part.
(Source BS 5304: 1988 'Safeguarding of machinery')

Entanglement hazards
The risk of the entanglement of clothing, hair and limbs can be present
wherever unguarded rotating shafts, drills or chucks are in operation.
(See Figure 21.)

Contact hazards
Contact with certain fixed and moving parts of machinery may cause
injuries, eg with a hot surface, belt fastenings on a moving conveyor or
the rough surfaces of a belt sander.

Ejection hazards
Some machines actually emit particles of metal. Grinding machines are
a classic example, where there is a risk of ejection of metal particles,

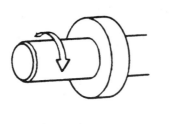 

Figure 21   Entanglement risks
(*Source* BS 5304: 1988 'Safeguarding of machinery')

particularly into the face and eyes or, alternatively, of the abrasive wheel actually bursting without warning.

Other hazards associated with machinery operation

While the above items are classed as the principal machinery hazards, there are many other hazards which need consideration at the design stage of a machinery-based work system, and during the operation of machinery. In certain cases people not actually involved in machinery operation may be exposed to risk of injury. Any assessment of the relative safety of a machine should take into account the following:

(a) procedures for job loading and removal;
(b) systems for changing of tools;
(c) safe removal of scrap and waste material;
(d) procedures for routine maintenance and adjustment, gauging, trying out following adjustment or setting, and in the event of breakdown;
(e) the potential for unexpected start-up or movement, uncovenanted stroke of the machine or mechanical failure;
(f) safe access to and egress from the machine and machine area;
(g) cleaning and housekeeping procedures in the machine area;
(h) availability of operating space; and
(i) potential risks to other persons passing through the machinery operating area.

Machinery guards

BS 5304 'Safeguarding of machinery' is the authoritative guidance on machinery safety in the United Kingdom, and is revised at regular intervals. The following forms of guard are specified in BS 5304.

Fixed guard

This guard has no moving parts associated with it, or dependent upon the mechanism of any machinery, and, when in position, prevents access to a danger point or area. Generally, a fixed guard should not be removable other than through use of a hand tool. Fixed guards are generally used to guard transmission machinery, such as belt drives to machinery.

Adjustable guard

This is a guard which incorporates an adjustable element which, once adjusted, remains in that position during a particular operation. Adjustable guards are commonly used with band saws.

Distance guard

A distance guard does not completely enclose a danger point or area but places it out of normal reach. Tunnel guards to metal cutting machines are a typical example.

Interlocking guard

This is a guard which has a movable part so connected with the machinery controls that:

(a) the part(s) of the machinery causing danger cannot be set in motion until the guard is closed;
(b) the power is switched off and the motion braked before the guard can be opened sufficiently to allow access to the dangerous parts; and
(c) access to the danger point or area is denied while the danger exists.

Interlocking guard systems may take a number of forms, ie mechanical, electrical, hydraulic, pneumatic or a combination of these forms. Such guards should 'fail to safety'. Failure to safety (fail-safe) is defined in BS 5304 as implying that any failure in, or interruption of, the power supply will result in the prompt stopping or, where appropriate, stopping and reversal of the movement of the dangerous parts before injury can occur, or the safeguard remaining in position to prevent access to the danger point or area.

Automatic guard

This guard is associated with, and dependent upon, the mechanism of the machinery and operates so as to remove physically from the danger area any part of a person exposed to the danger. This type of guard is commonly used with power presses. Certain automatic guards may be self-adjusting in that they prevent accidental access of a person to a danger point or area but allow the access of a workpiece which itself acts as part of the guard, the guard automatically returning to its closed position when the operation is completed. Portable circular saws frequently incorporate this form of self-adjusting automatic guard.

Machinery safety devices

A safety device is a protective appliance, other than a guard, which eliminates or reduces danger before access to a danger point or area can be achieved. Such devices take a number of forms.

Trip devices

This is a means whereby an approach by a person beyond the safe limit of working machinery causes the device to actuate and stop the machinery or reverse its motion, thus preventing or minimising injury at the danger point (BS 5304). Trip devices can operate on a mechanical, photo-electric or ultrasonic basis, or through pressure-sensitive systems, such as pressure-sensitive mats, which stop a machine immediately a person approaches the danger area.

Two-hand control devices

This is a device which requires both hands to operate the machinery controls, thus affording a measure of protection from danger to the operator. Such devices are featured in clicking presses in the footwear industry.

Overrun devices

A device which, used in conjunction with a guard, is designed to prevent access to machinery parts which are moving by their own inertia after the power supply has been interrupted, so as to prevent danger. Overrun devices may take the form of rotation sensing devices, timing devices and certain forms of braking system.

Mechanical restraint device

A device which applies mechanical restraint to a dangerous piece of machinery which has been set in motion owing to failure of the machinery controls or other parts of the machinery, so as to prevent danger. Such devices are commonly used on pressure die-casting machines.

Planned maintenance

Regulation 6 of PUWER requires that all work equipment shall be maintained 'in an efficient state, in efficient working order and in good repair'. Compliance with this absolute duty implies the operation of planned maintenance programmes. 'Planned maintenance' is defined in BS 3811: 1974 as; 'maintenance organised and carried out with forethought, control and the use of records to a predetermined plan'.

A planned maintenance programme should incorporate records which identify each item of work equipment, generally by serial number, the maintenance procedure to be undertaken, the frequency of maintenance and the individual responsible for ensuring the procedure is completed satisfactorily.

Machinery safety assessment

In assessing the relative safety of machines, the following form of audit should be undertaken.

1. Are machine controls correctly designed, safely located and clearly identified?
2. Is there an efficient stopping device for use in the event of an emergency? Is this device clearly identified?
3. Do the guards prevent access to danger points/areas? Do the safety devices operate effectively?
4. Do the features of the guarding system fail to safety? Are they 100 per cent reliable?
5. Can the safeguard be misused or defeated?
6. Does all electrical equipment comply with BS 2771?
7. Are there clearly identified access points for maintenance, lubrication and inspection?
8. What are the foreseeable machine failures? Does the guarding system cope effectively with such failures?
9. Is the machine correctly located so that other workers are not exposed to risk of injury caused by congestion of the working area?
10. Does the machine present secondary risks, such as noise or chemical hazards?
11. Does the machine create environmental pollution in the workplace, eg dust, fumes, gases? What controls are fitted to prevent such emissions?
12. Are temperature, lighting and ventilation levels in the machining area adequate and well maintained?

Summary

1. There is a legal duty on the employer, both under the HASAWA and PUWER, to provide safe machinery and plant.

2. Assessment of machinery safety must take into account mechanical factors and human factors.

3. The principal machinery hazards are associated with traps, entanglement, contact and ejection risks.

4. Secondary hazards can be created by failure to take into account the total system for machine operation.

5. BS 5304 'Safeguarding of machinery' provides authoritative guidance on all aspects of machinery safety.

6. Machinery guards and safety devices should be properly maintained at all times.

7. New machinery and plant should be assessed for compliance with BS 5304. Manufacturers, suppliers and importers of machinery and plant have clearly identified duties under Section 6 of the HASAWA.

References

British Standards Institution (1988) *BS 5304 Safeguarding of machinery* BSI, Milton Keynes

Health and Safety Executive *Guidance Note Series on Plant and Machinery* HMSO, London

Health and Safety Executive (1992) *Provision and Use of Work Equipment Regulations 1992 and Guidance on Regulations* HMSO, London

Stranks, J (1995) *Safety Technology* Pitman, London

Chapter 12
Fire Prevention

Fire is one of the natural elements, and generally taken to mean 'a state of burning or combustion'. Losses to industry through fire every year are substantial and, in order to prevent the deaths and human injury, damage to property and the consequent losses, it is essential that all employees at all levels in organisations are familiar with the causes of fire, fire protection procedures and the dangers associated with flammable substances.

The fire triangle

For combustion to take place and continue, three basic requirements are necessary. These are:

(a) fuel, in solid, liquid or gaseous form;
(b) an ignition source; and
(c) air.

These three components of fire are shown in the Fire Triangle (see Figure 22).

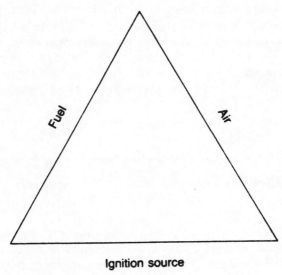

Figure 22 The fire triangle

If one of these three components is removed or not present, combustion cannot take place. Fire prevention strategies are, therefore, based on this fact, ie:

(a) limitation of, and control over, flammable substances stored on site;
(b) elimination of potential ignition sources; and
(c) restrictions on air available for combustion.

Causes of fire spread

The two principal areas where fire can spread are production areas and storage areas. In production areas will be found electrical equipment, processing plant and equipment. Frictional heat and sparks from these items can set fire to packaging materials, waste materials, goods being processed and even dust in substantial quantities. In storage areas, which tend to be sparsely occupied, fire spread can be caused as a result of smoking, through intruders – particularly children – committing acts of arson, or through defective electrical equipment setting fire to stored goods, packaging materials and certain combustible elements of buildings, such as wooden floors.

The principal causes of fire spreading in buildings are:

(a) lack of fire-separating walls between specific areas;
(b) poor housekeeping, resulting in combustible waste being stored in production and storage areas;
(c) oils and fats, which flow when in a burning state;
(d) the presence of vertical and horizontal features, such as trunking, ducting, conveyors, lift shafts and staircases, all of which readily permit fire spread from one part to another;
(e) the presence of dust at high level, which can explode in certain concentrations in air or, alternatively, burn rapidly once alight; and
(f) fires spreading from production to stored items in the same area.

Fire extinction

The following are the three principles of fire extinction.

Cooling

Cooling is the most commonly used method of putting out fires, using water. The application of water to a fire results in much of this water being converted to steam, thereby preventing much of the heat present, which would have supported combustion, being returned to the combustible material. A point is reached where the temperature drops to such an extent that the continuous ignition of combustible materials ceases. Water in spray form is the most efficient means of application to burning materials, as opposed to water jets or applying buckets of water to a fire.

Smothering

Smothering a fire, by preventing more oxygen flowing to the fire or by applying an inert gas to the fire, brings about a reduction in the air available for combustion to the point where no further combustion will take place.

Starvation

Starvation involves reducing the amount of fuel available for combustion. This is undertaken by removing the fuel from the fire, isolating the fire from the fuel source, and by reducing the quantity or bulk of fuel present.

Fire appliances

There are five principal types of hand-held fire appliance. Their use will depend upon the type of fire to be tackled. The two principal classes of fire and the appropriate fire appliances are shown in Table 3.

Class of fire	Description	Appropriate extinguisher
A	Solid materials, usually organic, with glowing embers	Water, foam, dry powder, vapourising liquid, carbon dioxide
B	Liquids and liquefiable solids: (i) miscible with water, eg methanol, acetone	Water, foam (but must be stable on miscible solvents) carbon dioxide, dry powder
	(ii) immiscible with water	Foam, dry powder, carbon dioxide, vapourising liquid

Table 3 Fire and fire appliance classification

Water appliances

There are two types of water appliance – stored pressure and gas cartridge. The stored pressure type contains carbon dioxide under pressure, water being expelled when the trigger is pulled. In the gas cartridge type, carbon dioxide is stored under pressure in a small cylinder. On breaking the seal with the plunger, the gas released expels the water through the nozzle.

Foam appliances

These are of the chemical foam type or stored pressure type. In both cases, when discharged and applied to burning material, they form a blanket of incombustible foam which prohibits further air for combustion. They are best used for small liquid spillage fires or for fires in small oil tanks whereby the foam can completely blanket the surface of the burning liquid. Foam appliances should always be completely discharged.

Carbon dioxide appliances

Carbon dioxide is suitable for application to both Class A and Class B fires, and fires involving electrical equipment. Liquid carbon dioxide is stored in a cylinder and, when discharged through a horn under its own pressure, quickly converts to carbon dioxide snow. The snow is subsequently converted to carbon dioxide gas by the fire. Carbon dioxide is not recommended for fires involving flammable liquids, unless the fire is very small.

Dry powder appliances

These appliances are of the stored pressure or gas cartridge type, and are extremely effective in dealing with flammable liquid fires.

Vapourising liquid appliances

These appliances incorporate a cylinder containing a liquid under pressure with dry carbon dioxide or nitrogen. On striking the knob, the seal is pierced thereby allowing the pressure to expel the liquid. A commonly used vapourising liquid is bromochlorodifluoromethane (BCF), which is particularly effective on fires involving electrical equipment and can be used quite satisfactorily for Class A and B fires, as well as small fires involving burning liquids.

Colour coding of fire appliances

All fire appliances are colour coded as in Table 4.

Extinguisher	*Colour code*
Water	Red
Foam	Cream
Carbon dioxide	Black
Dry chemical powder	Blue
Vapourising liquid	Green

Table 4   Colour coding of fire appliances

Fire protection systems

Such systems operate on the basis of detecting the presence of fire at a very early stage. They operate by three principal systems:

(a) heat sensing, where the actual temperature or rise in temperature is detected;
(b) smoke detection; and
(c) flame detection.

In all cases, these protection systems are linked with a form of fixed installation, such as a water sprinkler system or carbon dioxide system.

Legal requirements

Legal requirements are generally outlined in the Fire Precautions Act 1971 (FPA) and the Fire Safety and Safety of Places of Sport Act 1987.

Means of escape in case of fire

Means of escape in the case of fire is defined in the second-mentioned Act as 'escape from premises to some place of safety beyond the building, which constitutes or comprises the premises, and any area enclosed by it or within it'. It is important to distinguish means of escape as defined from an alternative means of escape, which is a second route, usually in the opposite direction, but which may join the first means of escape. It is not necessarily the second-best route.

In considering the establishment and maintenance of a means of escape a number of aspects must be considered. Discussions should take place with a representative of the local fire authority, eg the Fire Protection Officer, with a view to fire certification, where necessary, the designation of escape routes, the establishment of fire drills and fire instructions, and the procedures for sounding the fire alarm.

The following general requirements apply with regard to means of escape:

1. The total travel distance between any point in a building and the nearest final exit or protected stairway should not be more than:
   (a) 18 metres if there is only one exit; or
   (b) 45 metres if there is more than one exit.
   'Travel distance' is the distance a person has to travel from the fire to the commencement of the escape route. It should not be more than 50 metres, but in high-risk establishments may be as low as 10 metres. Where an alternative means is not available, the travel distance must be significantly reduced. The three stages of travel are

travel within rooms, travel within rooms to a stairway or final exit, and travel within stairways and to a final exit.

2. Two or more exits are necessary:
   (a) from any room in which more than 60 people work; or
   (b) if any point in the room is more than 12 metres from the nearest exit.

3. The minimum width of any exit should be 750 mm.

4. Corridors should not be less than one metre in width and in the case of offices, where corridors are longer than 45 metres, they should be subdivided by fire-resisting doors.

5. Stairways should be at least 800 mm in width, and should be fire-resistant, as should be connecting doors.

6. A single stairway is only sufficient in a building of up to four storeys.

7. The following are not acceptable as means of escape:
   (a) spiral staircases;
   (b) escalators;
   (c) lifts;
   (d) lowering lines; and
   (e) portable or throw-out ladders.

8. Fire doors must open outwards only.

9. Doors that provide means of escape must never be locked. In certain cases where they must be kept locked for security reasons, they should either be fitted with panic bolts or, alternatively, the key maintained in a key box close to the exit door. A suitable notice indicating this fact must be displayed adjacent to the exit.

10. A fire exit notice should be displayed above fire exit doors or attached to same.

11. Fire escape routes should be fitted with emergency lighting, together with signs indicating their purpose and the direction of travel to the exit.

12. Corridors and stairways forming a means of escape should be constructed of half-hour fire-resistant materials, and the surface finish should be non-combustible.

13. The fire alarm must be audible throughout the building.

14. Generally, no person should have to travel more than 30 metres to the nearest fire alarm point.

Fire instructions

Fire instructions are a notice advising occupants of a building of the action they must take on either hearing the fire alarm or discovering a fire. Such instructions must be displayed at prominent points in buildings.

Typical fire instructions are shown in Figure 23.

WHEN THE FIRE ALARM SOUNDS

1. Close the windows, switch off electrical equipment and leave the room, closing the door behind you.
2. Walk quickly along the escape route to the open air.
3. Report to the fire warden at your assembly point.
4. Do not attempt to re-enter the building.

WHEN YOU FIND A FIRE

1. Raise the alarm by . . . (If the telephone is to be used, the notice must include a reference to name and location.)
2. Leave the room, closing the door behind you.
3. Leave the building by the escape route.
4. Report to the fire warden at the assembly point.
5. Do not attempt to re-enter the building.

Figure 23   Fire instructions notice

Fire drills

A fire drill, resulting in total evacuation of the building, should take place at least annually. Fire wardens or other designated persons should take a head count on evacuation. Clearly identifiable assembly points should be established, and all persons, including visitors, should be aware of their specific assembly point.

Sounding the fire alarm

The fire alarm should be sounded weekly so that all occupants of a premises are familiar with its sound.

Summary

1. Fire is the principal cause of many deaths and substantial losses in British industry.

2. Everyone should be aware of the causes of fire and the ways that fire spreads in buildings.

3. The principal effects of fire appliances are cooling, smothering and starvation.

4. Staff should receive regular instruction in the selection and correct

use of fire appliances, although principal emphasis must always be on evacuating a building in the event of fire. Everyone should know the colour coding of fire appliances.

5. Means of escape in the event of fire should be clearly established and indicated.

6. Fire instruction notices must be displayed at strategic points in workplaces.

7. Fire drills must be undertaken at least annually, and fire alarms sounded weekly.

References

Dewis, M and Stranks, J (1988) *Fire Prevention and Regulations Handbook* RoSPA, Birmingham

Fire Precautions Act 1971, HMSO, London

Fire Precautions Act 1971 (Modifications) Regulations 1976 (SI 1976 No 2007), HMSO, London

Fire Offices Committee (1973) *Classification of Fire Hazards in Industry* FOC, London

Fire Protection Association (1982) *Fire, Safety and Security Planning in Industry and Commerce* Fire Data Sheet MR2, FPA, London

Fire Safety and Safety of Places of Sport Act 1987, HMSO, London

Home Office (1977) *Guide to the Fire Precautions Act 1971* HMSO, London

Home Office *Manual of Firemanship* HMSO, London

Lyons, W A (1981) *Action Against Fire* Alan Osborne & Associates, London

Chapter 13
Electrical Safety

Principles of electrical safety

The two principal hazards associated with the use of electricity are the risk of electrocution and shock. Fatal accidents associated with the use or misuse of electricity average at 50 per annum with approximately 1000 people being injured at work through shock and burns.

Legal requirements

The principal legal requirements relating to the safe use of electricity are the Electricity at Work Regulations 1989. Further guidance is given in the Memorandum of Guidance accompanying the Regulations, a number of British Standards and the Regulations for Electrical Installations published by the Institution of Electrical Engineers (the IEE 'Wiring Regulations').

These Regulations are made under the HASAWA. They impose duties on employers, the self-employed, eg electrical contractors, and employees ('duty holders') all of whom have equal levels of duty. The Regulations apply to all places of work and cover all electrical equipment ranging from battery-operated equipment to high voltage installations. They are framed in fairly general terms, which lay down principles of electrical safety, as opposed to the former Regulations which went into great detail. Broadly, they require precautions to be taken against the risk of death or personal injury from electricity in work activities, and cover systems, work activities and protective equipment, the strength and capability of electrical equipment, work in adverse or hazardous environments and the basic principles – insulation, protection and placing of conductors, earthing or other suitable precautions, connections, means for protecting from excess current, for cutting off the supply and for isolation, work on equipment made dead, work on or near live conductors, and the provision of working space, access and lighting.

Reg 16 requires persons to be 'competent to prevent danger and injury', implying the need for the appointment by management of competent persons (see Chapter 4) to cover electrical work activities. The defence of 'all reasonable precautions and all due diligence' (see Chapter 1) is available to a person charged with an offence under the Regulations.

Electrical safety, therefore, is primarily concerned with protecting people from electric shock, which could have fatal results, and from fire and burns arising from contact with electricity. Principal protective measures against electric shock are:

(a) protection against direct contact, ie by providing proper insulation for those parts of equipment liable to be electrically charged; and

(b) protection against indirect contact, eg by the provision of effective earthing for metallic enclosures which are liable to be charged with electricity if the basic insulation fails for any reason.

The following aspects are of significance in providing protection against electrocution and electric shock.

Earthing

Earthing implies connection to the general mass of earth in such a manner as will ensure at all times an immediate discharge of electrical energy without danger. Earthing, to give protection against indirect contact with electricity, can be achieved in a number of ways, including the connection of extraneous conductive parts of premises (radiators, taps, water pipes) to the main earthing terminal of the electrical installation. This creates an equipotential zone and eliminates the risk of shock that could occur if a person touched two different parts of the metalwork liable to be charged, under earth fault conditions, at different voltages. Where an earth fault exists, such as when a live part touches an enclosed conductive part, eg metalwork, it is vital to ensure that the electrical supply is automatically disconnected.

This disconnection is brought about by the use of overcurrent devices, ie correctly rated fuses or circuit breakers, or by correctly rated and placed residual current devices. The maintenance of earth continuity is crucial.

Fuses

A fuse is basically a strip of metal of such size as would melt at a predetermined value of current flow. It is placed in the electrical circuit and, on melting, cuts off the current to that circuit. Fuses should be of a type and rating appropriate to the circuit and the appliance it protects.

Circuit breakers

This device incorporates a mechanism that trips a switch from the 'ON' to 'OFF' position if an excess current flows in the circuit. A circuit breaker should be of the type and rating for the circuit and appliance it protects.

Earth leakage circuit breakers (residual current circuit breakers)

Fuses and circuit breakers do not necessarily provide total protection against electric shock. Earth leakage circuit breakers (ELCBs) provide protection against earth leakage faults, particularly at those locations where effective earthing cannot necessarily be achieved.

Reduced voltage

Reduced voltage systems are another form of protection against electric shock, the most commonly used being the 110 volt centre point earthed system. In this system the secondary winding of the transformer providing the 110 volt supply is centre tapped to earth, thereby ensuring that at no part of the 110 volt circuit can the voltage to earth exceed 55 volts.

Safe systems of work

Where work is to be undertaken on electrical apparatus or a part of a circuit, a formally operated safe system of work should be used. This normally takes the form of a permit to work system which ensures the following procedures:

(a) switching out and locking off of the electricity supply, ie isolation;
(b) checking by use of an appropriate voltage detection instrument that the circuit or part of same to be worked on is dead before work commences;
(c) high levels of supervision and control to ensure the work is carried out correctly;
(d) physical precautions, such as the erection of barriers to restrict access to the area, are implemented; and
(e) formal cancellation of the permit to work once the work is completed and return to service of the plant or system in question.

Causes and effects of shock

Electric shock results from an electric current flowing through the body. This has a direct effect on body organs and the central nervous system. The effect can be fatal if the heart rhythm is disturbed for long enough to stop blood flowing to the brain. Emergency action is vital in such cases. Resuscitation, ie mouth-to-mouth resuscitation, must be commenced quickly and be maintained until the patient recovers or, alternatively, is pronounced dead by a registered medical practitioner. The extent of the shock received is partly dependent upon the voltage of the current. However, the amperage is more significant than voltage, and an alternating current is more dangerous than a direct one. A current received through dry clothing is less dangerous than one received through wet clothing or directly on the bare skin and, obviously, all currents are more dangerous when the body is earthed than when insulation is provided by rubber-soled shoes or a rubber mat.

While the effects of electric shock can be fatal in many cases, non-fatal effects can include fractured bones and damage to flesh. For instance, the flesh at the point of entry may be damaged to a degree which ranges from a mild burn to severe destruction of muscles and internal organs. These injuries take a long time to heal because the mass of dead tissue takes a certain amount of time to separate itself from the living tissue. Sometimes the muscular spasm makes it impossible for the individual to let go of the object producing the shock and it passes for a longer period producing damage to the brain or paralysis of the heart or respiration system. Since a state of suspended animation may last for some time, it is important that resuscitation be carried on, in many cases, for hours following receipt of the electric shock.

Summary

1. The risk of electrocution and electric shock are the main hazards associated with electricity, although, in lesser cases, there can be severe body burns, muscular damage and fractured bones.

2. Electricity, when improperly used, can also be a cause of fires.

3. Protection strategies are directed at protection against direct contact, through the provision of proper insulation, and the avoidance of indirect contact, by good standards of earthing.

4. Other important strategies for protecting people against electric shock include the use of correctly rated fuses, circuit breakers, reduced voltage and the implementation of safe systems of work, in particular permit to work systems for work on electrical systems.

5. The effects of electric current flowing through the body should be understood by employees, together with procedures for resuscitation.

References

Beckingsale, A A (1976) *The Safe Use of Electricity* RoSPA, Birmingham

Electricity at Work Regulations 1989 (SI 1989 No 635), HMSO, London

Health and Safety Executive (1980) *Electrical Testing: Safety in Electrical Testing* (HSE Booklet HS(G) 13), HMSO, London

Health and Safety Executive (1989) *Memorandum of Guidance on the Electricity at Work Regulations, 1989* HMSO, London

Health and Safety Executive (1991) *Guidance for Small Businesses on Electricity at Work* HSE Information Centre, Sheffield

Health and Safety Executive *Electricity Regulations* (Booklet SHW 928), HMSO, London

Hughes, E (1978) *Electrical Technology* Longman, London

Imperial College of Science and Technology (1976) *Safety Precautions in the Use of Electrical Equipment* ICST, London

Chapter 14
Structural Safety

A substantial number of accidents in the workplace are associated, both directly and indirectly, with structural features – ie floors, stairs, floor openings, entrance and exit points. Poor standards of structural safety can result in slips and falls, some of which can result in major injuries like fractured arms and legs, and even death in certain cases. Good design of the structural aspects of working areas together with regular preventive maintenance are, therefore, crucial to providing a safe workplace.

The Workplace (Health, Safety and Welfare) Regulations 1992 (the 'Workplace' Regulations), and the supporting ACOP and HSE Guidance, lay down the principal requirements relating to structural safety. As with the MHSWR and PUWER, the majority of these duties on employers under the Workplace Regulations are of an absolute nature.

Floors and traffic routes

Floors and traffic routes must be of sound construction. They must not be sloped, uneven or slippery. For cleaning and other tasks using water, floors must be effectively drained. So far as is reasonable, every floor and traffic route must be kept free from obstruction and from articles or substances which could cause people to slip, trip or fall. Sufficient handrails and/or guards must be provided on staircases. (Regulation 12)

Falls or falling objects

Regulation 13 requires that suitable and effective measures be taken to prevent people from

(a) falling a distance likely to cause injury; and
(b) being struck by a falling object likely to cause personal injury.

Tanks, pits or structures must be securely covered or fenced to avoid the risk of people falling into a dangerous substance.

Windows, and transparent or translucent doors, gates and walls

These must be constructed of safety material. They must also be appropriately marked, or incorporate design features to make them easily apparent. (Regulation 14)

Windows, skylights and ventilators

No operable window, skylight or ventilator should be opened, closed or adjusted in a manner which causes any person to risk his health or safety. Similarly, when opened, they must not be in a position which is likely to expose a person to risk, eg of falling out of the window. (Regulation 15)

All windows and skylights must be designed, constructed and positioned so that they may be cleaned safely. (Regulation 16)

Organisation of traffic routes

There is a general duty to ensure the workplace is organised in such a way that pedestrians and vehicles can circulate in a safe manner. Traffic routes must be suitable – in size, number and position – for the persons or vehicles using them. All traffic routes must be well indicated. (Regulation 17)

Doors and gates

Doors and gates must be well constructed (including being fitted with any necessary safety devices). (Regulation 18)

Escalators and moving walkways

Escalators and moving walkways must:

(a) function safely;
(b) be equipped with any necessary safety devices; and
(c) be fitted with one or more emergency stop controls which are easily identifiable and readily accessible.

Access to and egress from the workplace

Section 2 of the HASAWA requires the occupier of the workplace to provide for his employees 'means of access to and egress from the workplace that are, so far as is reasonably practicable, safe and without risks to health'. This requirement applies to all employment situations, be the workplace an office block, factory, offshore platform or a coal face several miles below ground.

Typical risks associated with failure to comply with this requirement include defective and dangerous floors, staircases, catwalks, ladders, scaffolds and roadways, all of which present some form of risk to employees, visitors and members of the public who may use a premises such as a supermarket or public building.

Traffic systems

Poorly designed traffic systems can represent a serious hazard to employees and other persons visiting a premises. Well-controlled traffic systems around a workplace are, therefore, extremely important.

Features of traffic systems that are aimed at reducing the risk of personal injury, property damage and vehicular accidents, include the following:

(a) segregation of vehicular traffic routes from pedestrian routes;
(b) installation and operation of pedestrian barriers, pedestrian crossings and traffic lights at dangerous junctions;
(c) clear identification of vehicle parking areas, separating commercial vehicles from private cars, with directional signs at strategic locations;
(d) effective speed control over all vehicles, such as a 10 mph speed limit that is rigidly enforced and supported by 'sleeping policemen' or 'speed humps' which force traffic to slow down;
(e) operation of one-way systems where possible;
(f) control over unauthorised parking, parking in non-parking areas and generally 'unsafe parking' activities;
(g) the use of convex mirrors, located at strategic points, particularly where fork-lift traffic may cross recognised vehicle routes in or around the premises; and
(h) the provision and maintenance of high levels of artificial lighting at external access and egress points, loading bays, pedestrian walkways and parking areas. (It should be noted that good standards of artificial lighting can contribute greatly to improving site security.)

Summary

1. Good standards of structural safety are essential for the prevention of accidents associated with slips, trips and falls.

2. Special attention must be paid to the use of underground rooms used other than for storage purposes.

3. The provision of safe access to and egress from the workplace is a legal requirement.

4. A well-organised traffic system is essential for the prevention of accidents involving people and vehicles.

5. External lighting is an important aid to the maintenance of good traffic systems and in the prevention of breaches of site security, including theft from the premises and from vehicles parked on the premises.

References

Engineering Equipment Users Association (1973) *Factory Stairways, Ladders and Handrails* EEUA Handbook No 7, EEUA, London

Health and Safety Commission (1992) *Workplace (Health, Safety and Welfare) Regulations 1992 and Approved Code of Practice* HMSO, London

Health and Safety Executive (1978) *Road Transport in Factories* HMSO, London

Health and Safety Executive (1982) *Transport Kills* HMSO, London

Chapter 15
Construction and Contractors

The construction industry covers a wide field of operations, from the very large civil engineering projects such as motorway construction to, at the other end of the spectrum, self-employed tradesmen carrying out minor improvements and repairs to buildings. The extensive employment of casual labour, the relationship between the occupier of premises, contractors and sub-contractors, and the very real problems that can arise when contractors are working on an existing site, contribute to the high incidence of accidents in this industry. It is essential, therefore, for companies to establish clearly written procedures to regulate the activities of contractors when carrying out work on their behalf.

Construction hazards

The principal hazards in construction operations are associated with:

- falls from ladders, working platforms, pitched roofs;
- falls through fragile roofs, openings in flat roofs and floors;
- falls of materials;
- collapses of excavations;
- site transport activities;
- machinery and powered hand tools;
- housekeeping failures;
- fire;
- personal protective equipment.

Recent statistics (1993–94) on fatal accidents in construction are:

Falls from a height	56%
Trapped by something collapsing or overturning	21%
Struck by a moving vehicle	10%
Contact with electricity or an electrical discharge	5%
Struck by falling/flying object during machine lifting of materials	4%
Contact with moving machinery or material being machined	3%
Exposure to a hot or harmful substance	1%

Contractors and the law

Contracting activities are covered under both civil law and criminal law. Under civil law, ie the law giving rise to claims for damages, contractors employed by a company are at risk of civil liability in damages from their own employees and other people, including company employees, if they fail to take adequate care for their safety during their operations on company premises and such persons are injured as a result.

Under criminal law, eg HASAWA, the prime responsibility rests with the occupier of the premises or site concerned. The company, therefore, is subject to prosecution if there are breaches of statutory provisions on its premises if these breaches are caused by the contractor or one of his sub-contractors. On this basis a company must take all reasonably practicable measures to ensure compliance with these provisions throughout its premises, ie they must make such arrangements which, if implemented, would ensure legal compliance, and then ensure that they are carried out. Moreover, if the making of such arrangements by the company is not reasonably practicable, due to the nature of the contractor's activities, the company must do its best to ensure that the contractor makes appropriate arrangements and then carries them out.

The legal provisions relating to safety, health and welfare in construction activities are outlined in:

- Construction (Lifting Operations) Regulations 1961
- Construction (Head Protection) Regulations 1989
- Construction (Design and Management) (CDM) Regulations 1994
- Construction (Health, Safety and Welfare) Regulations 1996

The general provisions of the HASAWA apply to all construction-related activities. In particular, contractors have obligations under Section 3 (duties of employers to other than their employees), and may have obligations under Section 4 not to endanger persons who are not their employees. In particular, duties under the CDM Regulations must be read in conjunction with the duties under the Management of Health and Safety at Work Regulations 1992.

Head protection

The Construction (Head Protection) Regulations 1989 impose duties on employers, people in control of construction sites, the self-employed and employees regarding the provision and use of head protection wherever there is a reasonably foreseeable risk of head injury. Employers must provide and maintain head protection, and ensure it is worn wherever there is risk of head injury. Persons in control of construction sites, eg main contractors, site managers, in conjunction with employers must identify when and where head protection must be worn, inform operators accordingly, and supervise the wearing of same.

Employees, similarly, have a duty to wear head protection in designated areas and operations.

The Construction (Design and Management) (CDM) Regulations 1994

The Construction (Design and Management) (CDM) Regulations 1994, specify the relationships which must exist between a client, principal contractor and other contractors from a health and safety viewpoint. Under these regulations a number of people have both general and specific duties.

Thus a *client*, namely a person for whom a project is carried out, must:

(a) appoint a planning supervisor and a principal contractor in respect of each project;
(b) ensure that the planning supervisor has been provided with information about the state or condition of specified premises; and
(c) ensure that information in a health and safety file is available for the inspection of specified persons.

A *planning supervisor* must ensure that the HSE is informed of the specified particulars of a notifiable project. (Particulars to be provided for the HSE are detailed in Schedule 1 of the regulations.) A notifiable project is one where the construction phase:

(a) will be longer than 30 days; or
(b) will involve more than 500 person days of construction work.

Planning supervisors have specific duties in respect of:

(a) the design of any structure comprised in a project;
(b) the co-operation between designers;
(c) the giving of adequate advice to specified persons;
(d) the preparation, review and necessary amendment of a health and safety file; and
(e) the delivery of the health and safety file to the client.

The *principal contractor* has a number of duties in respect of:

(a) co-operation between contractors;
(b) compliance with the health and safety plan;
(c) the exclusion of unauthorised persons;

(d) the display of notices;
(e) the provision of information to the planning supervisor;
(f) the provision of certain health and safety information to contractors, and the provision of specified information and training to the employees of those contractors;
(g) ensuring that the views and advice of persons at work on the project, or their representatives, concerning matters relating to their health and safety are received, discussed and co-ordinated.

The principal contractor is empowered, for certain purposes, to give directions to contractors and to include rules in the health and safety plan.

In the case of a *designer*, he:

(a) is prohibited from preparing a design unless the client for the project is aware of his duties under the regulations and of the requirements of any practical guidance issued by the HSC;
(b) must ensure that the design he prepares, and which is to be used for the purposes of construction work or cleaning work, takes into account among design considerations certain specified matters.

A *contractor* (other than the principal contractor) must co-operate with the principal contractor in order to enable him to comply with his duties as indicated above.

A person who appoints a planning supervisor, or arranges for a designer to prepare a design, or for a contractor to carry out or manage construction work, is prohibited from doing so unless he is reasonably satisfied:

(a) as to the competence of those appointed or arranged; and
(b) as to the adequacy of the resources allocated or to be allocated for the purposes of performing their respective functions by those appointed or arranged.

Both the planning supervisor and principal contractor must comply with the requirements relating to the health and safety plan, and the commencement of the construction phase of a project is prohibited unless a health and safety plan has been prepared in respect of the project.

Competent persons
No person shall be appointed by a client as his agent unless the client is reasonably satisfied as to that person's competence to perform the duties imposed on the client under the regulations.

Similarly, planning supervisors, designers and contractors must be competent in terms of complying with any requirements and conducting his undertaking without contravening any prohibitions imposed by or under health and safety law.

Health and safety files

The following information must be incorporated in a health and safety file:

(a) information included with the design which will enable the planning supervisor and designers to comply with legal requirements; and

(b) any other information relating to the project which is considered necessary to ensure the health and safety of all persons involved in the project.

Health and safety plans

An important requirement is that the planning supervisor shall ensure that a health and safety plan for the project has been prepared in sufficient time for it to be provided to a contractor before arrangements are made to carry out or manage the construction work.

Under Regulation 15, the following information must be incorporated in a health and safety plan:

(a) a general description of the construction work comprised in the project;

(b) details of the time in which it is intended that the project, and any intermediate stages, will be completed;

(c) details of the risks to the health or safety of any person carrying out the construction work so far as such risks are known to the planning supervisor or are reasonably foreseeable;

(d) any other information concerning the competence of the planning supervisor and the availability of resources to enable him to undertake his specified duties;

(e) information held by the planning supervisor and needed by the principal contractor to enable him to comply with his specified duties; and

(f) information held by the planning supervisor and which any contractor should know in order for him to comply with statutory provisions in respect of welfare.

The principal contractor must then take suitable measures to ensure that the health and safety plan incorporates appropriate features for ensuring the health and safety of all persons involved.

Construction (Health, Safety and Welfare) Regulations 1996

The regulations impose requirements with respect to the health, safety and welfare of persons at work carrying out construction work as defined and of others who may be affected by that work. Specified regulations apply in respect of construction work carried out on a construction site as defined, and where a workplace on a construction site is set aside for purposes other than construction, the regulations do not apply.

Subject to specific exceptions, the regulations impose requirements on employers, the self-employed and others who control the way in which construction work is carried out (duty holders). Employees have duties in respect of their own actions. Every person at work has duties as regards co-operation with others and the reporting of danger.

The principal duties of employers, the self-employed and controllers

Safe places of work
A general duty to ensure a safe place of work and safe means of access to and from that place of work.

Specific provisions include the following:

Precautions against falls

(a) Prevention of falls from heights by physical precautions or, where this is not possible, provision of equipment that will arrest falls.
(b) Provision and maintenance of physical precautions to prevent falls through fragile materials.
(c) Erection of scaffolding, access equipment, harnesses and nets under the supervision of a competent person.
(d) Specific criteria for using ladders.

Falling objects

(a) Where necessary to protect people at work and others, taking steps to prevent materials or objects from falling.
(b) Where it is not reasonably practicable to prevent falling materials, taking precautions to prevent people from being struck, eg covered walkways.
(c) Prohibition of throwing any materials or objects down from a height if they could strike someone.
(d) Storage of materials and equipment safely.

Work on structures

(a) Prevention of accidental collapse of new or existing structures or those under construction.
(b) Ensuring any dismantling or demolition of any structure is planned and carried out in a safe manner under the supervision of a competent person.
(c) Only firing explosive charges after steps have been taken to ensure that no one is exposed to risk or injury from the explosion.

Excavations, cofferdams and caissons

(a) Prevention of the collapse of ground both in and above excavations.
(b) Identification and prevention of risk from underground cables and other services.
(c) Ensuring cofferdams and caissons are properly designed, constructed and maintained.

Prevention or avoidance of drowning

(a) Taking steps to prevent people falling into water or other liquid so far as is reasonably practicable.
(b) Ensuring that personal protective and rescue equipment is immediately available for use and maintained, in the event of a fall.
(c) Ensuring sure transport by water is under the control of a competent person.

Traffic routes, vehicles, doors and gates

(a) Ensuring construction sites are organised so that pedestrians and vehicles can both move safely and without risks to health.
(b) Ensuring routes are suitable and sufficient for the people or vehicles using them.
(c) Prevention or control of the unintended movement of any vehicle.
(d) Ensuring arrangements for giving a warning of any possible dangerous movement, eg reversing vehicles.
(e) Ensuring safe operation of vehicles including prohibition of riding or remaining in unsafe positions.
(f) Ensuring doors and gates which could prevent danger, eg trapping risk of powered doors and gates, are provided with suitable safeguards.

Prevention and control of emergencies

(a) Prevention of risk from fire, explosion, flooding and asphyxiation.
(b) Provision of emergency routes and exits.
(c) Provision of arrangements for dealing with emergencies, including procedures for evacuating the site.
(d) Where necessary, provision of fire-fighting equipment, fire detectors and alarm systems.

Welfare facilities

(a) Provision of sanitary and washing facilities and an adequate supply of drinking water.
(b) Provision of rest facilities, facilities to change and store clothing.

Site-wide issues

(a) Ensuring sufficient fresh or purified air is available at every workplace, and that associated plant is capable of giving visible or audible warning of failure.
(b) Ensuring a reasonable working temperature is maintained at indoor workplaces during working hours.
(c) Provision of facilities for protection against adverse weather conditions.
(d) Ensuring suitable and sufficient emergency lighting is available.
(e) Ensuring suitable and sufficient lighting is available, including the provision of secondary lighting where there would be a risk to health or safety if the primary or artificial lighting failed.
(f) Maintaining construction sites in good order and in a reasonable state of cleanliness.
(g) Ensuring the perimeter of a construction site to which people, other than those working on the site, could gain access, is marked by suitable signs so that its extent can be easily identified.
(h) Ensuring all plant and equipment used for construction work is safe, of sound construction and used and maintained so that it remains safe and without risks to health.

Training, inspection and reports

(a) Ensuring construction activities where training, technical knowledge or experience is necessary to reduce risks of injury are only carried out by people who meet these requirements or, if not, are supervised by those with appropriate training, knowledge or experience.
(b) Before work at height, on excavations, cofferdams or caissons

begins, ensuring the place of work is inspected, (and at subsequent specified periods), by a competent person, who must be satisfied that the work can be done safely.
(c) Following inspection, ensuring written reports are made by the competent person.

Accidents in construction activities

Falling from a height is the most common type of fatal accident in the construction industry. Manual handling accidents, frequently resulting in hernias, back injuries, foot and hand injuries, are the most common type of non-fatal accident.

Other frequent types of accident include:

(a) ladder accidents (the one out:four up rule for ladders should always be followed);
(b) falls from working platforms, such as scaffolds and mobile platforms;
(c) falls of materials, such as bricks, roof tiles, etc, on to people working below;
(d) falls from pitched roofs or through fragile roofs;
(e) falls through openings in flat roofs and floors;
(f) collapses of excavations, resulting in people being buried alive;
(g) accidents associated with site transport, eg dumper trucks, reversing lorries;
(h) use of machinery and powered hand tools; and
(i) accidents associated with bad housekeeping.

Regulating the contractor

Contractors undertake a wide range of services, from window cleaning to minor repairs and improvements to buildings to large-scale construction projects on existing or 'green field' sites. In the majority of situations the client and main contractor are jointly responsible for ensuring standards of safety, health and welfare are maintained during construction operations. Only in a 'green field' site situation, before new premises are handed over to the occupier, does the main contractor have full control and, therefore, responsibility, for a site.

The presence of contractors on site for a lengthy period of time can, if not carefully planned, create numerous problems, in addition to the risk of accidents to staff and site employees. To prevent such problems arising, therefore, it is essential that any project is discussed with the main contractor prior to commencement. In particular, the site of operations should be clearly defined on a factory or premises plan, indicating specific access and egress points for workers and materials, parking areas, vehicular traffic routes, including those of fork-lift trucks, areas prohibited to contractor's staff, the location of fire emergency points and specific hazards which may exist, such as the presence of flammable liquid storage points. Furthermore, agreement should be reached on the use or otherwise of company amenities by the contractor's staff, including washing, shower and toilet facilities, catering arrangements, first aid stations and, possibly, arrangements for drying clothing. The contractor should be advised of any company processes or activities which may expose building workers to health and safety risks.

It is normal practice for the contractor to take out insurance cover to indemnify the client in respect of any negligence resulting in personal injury and/or death, or damage to property and plant, arising out of or in connection with the work.

Other aspects which will need agreement between the client and contractor include:

(a) circumstances surrounding use of the owner's equipment;
(b) procedures for reporting, recording and investigating accidents on site;
(c) the use of safe systems of work, including permits to work;
(d) the safe use of plant and machinery, including electrical equipment and the supply of electricity to the site;
(e) the prevention of noise nuisance to the inhabitants of the neighbourhood;
(f) fire protection procedures, including the need for safe welding operations;
(g) arrangements for the safe removal of dangerous substances and wastes, in particular asbestos lagging or other asbestos-based materials which may be incorporated in the buildings due for redevelopment;
(h) the supply and use of appropriate personal protective equipment by site workers;
(i) permitted routes for contractors' vehicles, parking areas, delivery points, vehicle safety standards and driving restrictions;
(j) procedures for site clearance on completion; and
(k) the procedures for site security.

Such items should be incorporated in the 'Rules for the management of the construction work'. (See Regulation 16, CDM Regulations.)

Local management should also agree a form of safety monitoring covering contractor activities with the main contractor, including the right to dismiss from the site any contractor, sub-contractor or individual who, in the eyes of the company, has behaved unsafely to the extent of exposing himself, other site workers or members of the occupier's workforce to risk or injury or, alternatively, caused an accident to take place as a result of their negligence or lack of safety precautions.

Safety method statements

A safety method statement is a formally written safe system of work, or series of integrating safe systems of work, agreed between client and main contractor or main contractor and sub-contractor, and produced where work with a foreseeably high hazard content is to be undertaken.

It should specify the operations to be carried out on a stage-by-stage basis and indicate the precautions necessary to protect site operators, staff occupying the premises where the work is undertaken and members of the public who may be affected directly or indirectly by the work.

It may incorporate information and specific requirements stipulated by clients, employers, health and safety specialists, enforcement officers, the police, site surveyor and the manufacturers and suppliers of plant, equipment and substances used during the work. In certain cases, it may identify training needs, eg under the Control of Substances Hazardous to Health (COSHH) Regulations, Electricity at Work Regulations, or the use of competent persons or specially trained operators.

The use of safety method statements

A safety method statement is necessary to ensure safe working in activities involving:

(a) the use of hazardous substances in large quantities, eg toxic, corrosive, harmful, irritant, flammable, etc substances;
(b) the use of explosives;
(c) lifting operations;
(d) potential fire risk situations;
(e) electrical hazards;
(f) the use of sources of radiation;
(g) the risk of dust explosions or the inhalation of hazardous dusts, gases, vapours, fumes, etc;
(h) certain types of excavation, particularly those adjacent to existing buildings;
(i) demolition work; and
(j) the removal of asbestos from buildings.

Contents of the safety method statement

The following features may be incorporated in a safety method statement:

(a) techniques to be used;
(b) access provisions;

(c) safeguarding of existing work locations and positions;
(d) structural stability requirements, eg shoring;
(e) procedures to ensure the safety of others, eg members of the public;
(f) health precautions, including the use of local exhaust ventilation systems and personal protective equipment;
(g) plant and equipment to be used;
(h) procedures to prevent area pollution;
(i) segregation of certain areas;
(j) procedures for the disposal of toxic wastes; and
(k) procedures to ensure compliance with specific legislation, eg Environmental Protection Act, COSHH Regulations, Noise at Work Regulations, Control of Asbestos at Work Regulations.

Contractors' Regulations

Most large organisations who employ contractors of all types on a regular basis operate a form of Contractors' Regulations which lay down the ground rules for any organisation operating on their site. It is a condition of contract that contractors and sub-contractors comply with such Regulations at all times.

Maintenance work

Maintenance work may be of a planned and routine nature. Other forms of maintenance may be required in crisis situations.

The principal hazards associated with maintenance work can be classified:

1. **Mechanical:** Machinery traps, entanglement, contact, ejection; unexpected start-up of machinery.
2. **Electrical:** Electrocution, shock, burns, fire.
3. **Pressure:** Unexpected pressure releases, explosion.
4. **Physical:** Extremes of temperature, noise, vibration; dust and fumes.
5. **Chemical:** Gases, fogs, mists, fumes, etc.
6. **Structural:** Obstructions, floor openings.
7. **Access:** Work at heights, work in confined spaces.

The following precautions should be considered:

1. Safe systems of work.
2. Permit to work systems.
3. Designation of competent persons.
4. Use of method statements.
5. Operation of company Contractors' Regulations.

6. Controlled areas.
7. Access control.
8. Information, instruction and training.
9. Supervision arrangements.
10. Signs, marking and labelling.
11. Personal protective equipment.

Summary

1. The relationship between contractor, sub-contractor and client should be appreciated in terms of both criminal and civil liability in the event of accidents in construction activities.

2. Criminal provisions relating to safety, health and welfare on construction sites are detailed in the various Construction Regulations.

3. The principal type of construction accident, frequently with fatal results, is a fall from a height. Hernias, foot, hand and back injuries are common.

4. Under the CDM Regulations clients, planning supervisors, designers, principal contractors and contractors have specific responsibilities.

5. With construction projects, a health and safety plan must be prepared at the pre-tender and construction phases of the project.

6. The Construction (Health, Safety and Welfare) Regulations 1996 impose requirements with respect to the health, safety and welfare of persons carrying out construction work.

7. A conscious effort should be made by organisations to regulate the activities of contractors, commencing with careful pre-planning of the operation prior to work commencing.

8. Contractors should be required to produce and comply with safety method statements for identified highly hazardous activities.

9. All organisations should operate formal contractors' regulations aimed at establishing the ground rules for contracting activities of all types on their premises.

10. Maintenance work can expose employees and other persons at work to a range of hazards which should be carefully controlled by employers.

References

Anderson, P W P (1989) *Safety Manual for Mechanical Plant Construction* Kluwer, London
Armstrong, P T (1980) *Fundamentals of Construction Safety* Hutchinson, London

Building Regulations 1985 (SI 1985 No 1065), HMSO, London

Construction (Design and Management) Regulations 1994 HMSO, London

Construction (Health, Safety and Welfare) Regulations 1996 (SI 1996 No 1592), HMSO, London

Construction (Head Protection) Regulations 1989, HMSO, London

Construction (Lifting Operations) Regulations 1966 (SI 1966 No 1581), HMSO, London

Health and Safety Commission (1995) 'Approved Code of Practice: Managing Construction for Health and Safety', *Construction (Design and Management) Regulations 1994*, HSE Books, Sudbury

Health and Safety Executive (1995) *Guidance for Designers – Designing for Health and Safety in Construction*, HSE Books, Sudbury

Health and Safety Executive (1995) *Guidance for Clients, Planning Supervisors, Designers and Contractors – A Guide to Managing Health and Safety in Construction*, HSE Books, Sudbury

Health and Safety Executive (1995) *Guidance for Builders and Contractors – Health and Safety for Small Construction Sites*, HSE Books, Sudbury

Health and Safety Executive (undated) *Roofwork: Prevention of Falls* (Guidance Note GS 10), HMSO, London

Health and Safety Executive (undated) *Safe Erection of Structures; Pts I–IV* (Guidance Note GS 28/1–4), HMSO, London

Health and Safety Executive (undated) *Health and Safety in Demolition Work, Pts I–IV* (Guidance Note 19/1–4), HMSO, London

International Labour Organisation (1982) *Safety and Health in Building and Civil Engineering Work* ILO, Geneva

Royal Society for the Prevention of Accidents (undated) *The Supervisor's Guide to the Construction Regulations* RoSPA, Birmingham

Chapter 16
Mechanical Handling

The movement and handling of goods may be by manual or mechanical means. This chapter examines the various forms of mechanical handling and the safety requirements necessary for such operations. (The various aspects of manual handling operations are covered in Chapter 5.)

Clearly, wherever possible, mechanical systems should be used in preference to manual ones. The type and form of the handling system will vary considerably according to the form of the loads to be handled, their shape, size, weight, the distance of travel and frequency of movement. Factors, such as the form of storage system in operation, such as a drive-in pallet racking system, the general layout of the premises and external areas, and procedures for the receipt and despatch of raw materials and finished products respectively, will also determine the type of mechanical handling system used.

Mechanical handling systems

Generally, mechanical handling systems take the following three forms:

(a) conveyors, which move goods either on a horizontal plane or at inclined angles;
(b) elevators, which operate on a vertical or inclined angle basis; and
(c) mobile handling equipment, such as fork-lift trucks which, fundamentally, transfer loads from one point to another, both internally and externally.

Conveyors
Conveyors take a number of forms according to the materials they may be carrying – roller conveyors, slat conveyors, trough conveyors, chain conveyors, belt conveyors and screw conveyors. While individual forms of conveyor may incorporate hazards specific to themselves, the more general hazards associated with all types of conveyor are outlined below:

(a) 'nips' or traps between moving parts, eg between belt and drive, between driven and tension rollers (see Figure 24);
(b) traps between moving and fixed parts, eg between slats on a slat conveyor and the guide frame containing the conveyor;

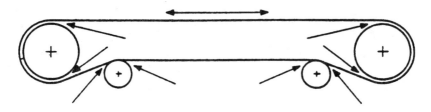

Figure 24 Trapping points on a reversible belt conveyor
(*Source* BS 5304: 1988 'Safeguarding of machinery')

(c) traps and nips created by a drive mechanism, eg between chains and sprockets on a chain conveyor (see Figure 25);

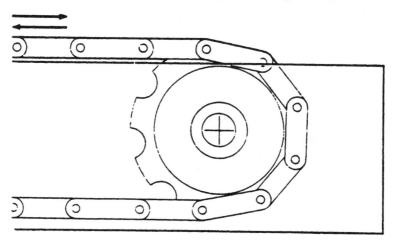

Figure 25 Traps between sprocket and chain on a reversible chain conveyor
(*Source* BS 5304: 1988 'Safeguarding of machinery')

(d) traps created at transfer points between two conveyors, eg between a roller conveyor and belt conveyor (see Figure 26);

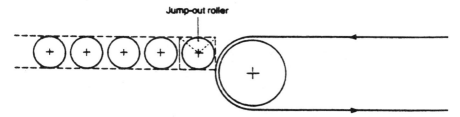

Figure 26 Safety requirements at the junction between a belt conveyor and roller conveyor. A removable or 'jump out' roller is installed at the end of the roller conveyor.
(*Source* BS 5304: 1988 'Safeguarding of machinery')

190

(e) hazards associated with sharp edges to, for instance, worn conveyor chains, which may be exposed.

Guarding of conveyors

Fixed guards, which totally enclose the trap or nip created, should always be used in preference to other forms of guard. However, where frequent access may be needed to a danger point, eg for lubrication, an interlocked guard, which automatically stops the conveyor when it is lifted, should be installed. In certain cases, a tunnel guard, which is a form of distance guard, may be appropriate. (See Chapter 11.) This form of guard prevents access to the danger point because of the relationship of the guard opening dimensions to the length of the tunnel. Clearance between the sides of the opening of a tunnel guard and the goods being conveyed should be not less than 50 mm.

As with all other types of machinery, conveyor guarding should comply with BS 5304 'Safeguarding of machinery', particularly with regard to any openings in fixed guards.

Further points for consideration in the design and operation of conveyors include:

(a) arrangements for lubrication with the guards in position;
(b) maximising the radius of all bends to prevent items carried jamming on the conveyor or falling off same;
(c) elimination of sharp edges to supporting structural members; and
(d) the fitting of rails or side members to conveyors where the conveyor rises to more than 1 metre above floor level.

Where a conveyor is longer than 20 metres, an emergency stop (trip) wire should be fitted or, alternatively, a series of emergency stop buttons, spaced at not more than 10 metre intervals, should be provided. Stop buttons should be mushroom-shaped and red. They should remain in the 'OFF' position until reset.

Elevators

An elevator is used to transfer goods either vertically, eg between the floors of a building, or at a specific angle, eg between one floor and another. They may also be of the movable type for general use in warehouses and storage areas.

Both ends of an elevator should incorporate fixed guards due to the traps created between elevator chains and sprockets. Bucket elevators are commonly used for transferring loose materials, such as grain, inside a fixed shaft or hoistway. In this case there is a significant risk of dust explosions and, to prevent such explosions, all elevator heads must be fitted with explosion reliefs, usually a form of blow-out panel of a minimum size according to the size of the elevator.

With bar elevators, used for transferring sacked goods or boxed goods, traps can be created between the bar and the fixed part of the elevator. Such elevators should be guarded at the top and bottom by fixed guards.

Mobile handling equipment

The most commonly used item of mobile handling equipment is the fork-lift truck, which can include pedestrian-operated stacking trucks, reach trucks, counterbalance fork trucks, narrow aisle trucks and order pickers. Where such equipment is used, it is necessary to consider certain aspects such as the safe operation and maintenance of the equipment, specific provisions relating to the area of operation, and particular requirements relating to operators in terms of training, supervision and control, general fitness and the use of personal protective equipment. These aspects are dealt with below.

Mechanical handling equipment
To ensure safe operation, the following rules should be:

(a) the maximum rated load capacity of the truck, as shown on the manufacturer's identification plate, should not be exceeded;
(b) passengers should never be carried, unless in a properly manufactured cage or platform;
(c) untrained and/or unauthorised personnel should never drive or operate such equipment;
(d) when driver-operated trucks are left unattended they should have the forks lowered and the truck should be immobilised by putting the controls in the neutral position, shutting off the power, applying the brakes, and removing the key or connector plug;
(e) the keys to the truck should be kept in a secure place when not in use;
(f) a formally established maintenance programme, based on the manufacturer's recommendations for inspection, maintenance and servicing, should be operated;
(g) drivers should be trained to undertake periodic maintenance checks, which should be linked to a formally established defect reporting system;
(h) trucks must comply with the Road Traffic Acts, in terms of lights, brakes, steering, etc, when operating on a public highway; and
(i) all equipment used for mechanical handling and lifting should be subject to a six-monthly and annual examinations. (Lifting chains should be inspected annually and certificated as such in accor-

dance with statutory requirements, ie all these items of equipment are classed as 'lifting machines' under section 27 of the FA, and the provisions relating to the examination and testing of same apply.

Area of operation

To ensure safe operation, consideration must be given to the design, layout and maintenance of operating areas. The following aspects must be considered:

(a) floors should be smooth, free from obstructions, changes in floor level and of adequate load-bearing capacity;

(b) sharp bends and overhead hazards, such as pipework, ducting and electrical conduits, should be eliminated;

(c) edges of loading bays should be protected when not in use, and tiger-striped floor markings incorporated at dangerous points to warn the driver;

(d) ramps and gradients should not exceed 10 per cent and there should be a gradual change of gradient at the bottom and the top of the slope;

(e) lighting should be maintained at a minimum level of 100 lux;

(f) bridge plates should be designed with an adequate safety margin to support loaded equipment;

(g) aisles should be wide enough, kept clear at all times and, in areas where other personnel are working on a spasmodic basis, fitted with photo-electrically operated audible warning devices, to warn such personnel of the presence of the truck;

(h) notices requiring drivers to sound their horns should be located at key points in the operating area, and corners should be fitted with convex mirrors to enable the driver to see people and goods prior to turning the corner;

(i) in truck charging areas, ventilation should be sufficient to prevent accumulations of hydrogen gas building up;

(j) refuelling of trucks should take place outside the area of operation; and

(k) truck parking areas should be located away from main vehicle parking areas.

Operator requirements

Operators of such equipment should be selected on the basis of their attitude to safe vehicle operation, physical fitness and attendance at an approved driver training course. Companies should operate a form of documentation of authorised operators through the use of a 'permit to drive' system. (An example of a company permit to drive is shown in Figure 27.)

PERMIT TO DRIVE

Name _____   Date of Birth _____

Department _____   Date of Basic Training _____
                              'Certificate No.' _____

Permit Number _____   Date of Issue _____

The person named in this Permit is authorised to drive the following powered materials handling trucks:-

CLASS:– A; B1; B2; C; D; and E. (delete as appropriate)

Signature of Permit Holder _____

Signature of Unit Manager _____   Date _____

CLASS OF TRUCK

A – pedestrian stacking;

B1 – pedestrian counterbalance;

B2 – rider counterbalance;

C – reach;

D – order picker;

E – narrow aisle.

NOTE

1 This permit does not authorise the person to whom it is issued to drive on a public road.

2 The permit is only valid for the class of truck specified.

3 This permit may be withdrawn if the safety rules laid down in the Alpha Company Code of Practice 'Materials Handling and Storage Equipment', are not observed.

Figure 27   Typical form of company permit to drive

Fork-lift truck handling operations, particularly in warehouses and stores where more than one truck may be in operation, require a high level of supervision and control, supported by regular training and retraining of operators. Operator training should be carried out by trainers who are experienced in the specific tasks to be undertaken.

Operator training should take place in the following three specific stages:

(a) acquisition of the specific skills and knowledge required to operate the equipment safely and to carry out the prescribed daily checks on the equipment;

(b) on-the-job training in a designated training area to develop operational skills; and

(c) familiarisation training, under close supervision of the trainer, in the workplace.

Further training should be carried out when:

(a) the driver is transferred to a new operational area or specific handling activity;

(b) new or modified equipment is introduced;

(c) there has been a significant change in the working layout; and

(d) there is evidence to indicate that driving standards have deteriorated or there has been a lapse in operator standards.

Operators should be provided with safety footwear and a safety helmet. In certain cases, the provision of hearing protection may be necessary. Protective clothing to suit weather and/or temperature conditions should also be provided where operations extend to external areas or into temperature-controlled parts of the premises, such as cold stores. In these cases donkey jackets and gloves should be provided.

A record should be maintained of all authorised operators. This record should indicate the:

(a) name of the operator;

(b) date of passing the truck driving test;

(c) serial number of the permit to drive;

(d) date of initial training and retraining; and

(e) classes of truck which the operator is authorised to drive.

The HSE publication *Rider-operated Lift Trucks – Operator Training: Approved Code of Practice and Supplementary Guidance* (1988) provides excellent guidance on this matter.

Summary

1. Conveyors can be exceedingly dangerous if not effectively guarded in accordance with BS 5304 and there is an absolute duty under PUWER to guard same.

2. Fixed guards should always be used on conveyor danger points wherever practicable, together with the provision of emergency stop devices.

3. Elevators should be guarded at both ends due to traps created by chains and sprockets (toothed wheels) in particular.

4. Where elevators are used for the transfer of dust-forming materials, explosion reliefs should be installed at the head and the hoistway/shaft and floor openings constructed in materials to give a half-hour notional or predicted period of fire resistance.

5. If they are not properly controlled, mechanical handling equipment, such as fork-lift trucks, can represent a serious risk to the workforce.

6. Special consideration must be given to the equipment, the design and layout of the working area, and to the selection, training, supervision and control of vehicle operators.

7. Clearly established maintenance procedures should be operated in the case of all mobile handling equipment.

8. A formal permit to drive system should be operated with a view to regulating the selection and appointment of truck drivers. Where there is evidence of unsafe driving or truck operation, consideration must be given to the withdrawal of the permit to drive as a form of disciplinary action.

References

British Standards Institution (1988) *BS 5304, Safeguarding of machinery* BSI, Milton Keynes

Health and Safety Executive (1979) *Lift trucks* HMSO, London

Health and Safety Executive (1980) *Safe Working with Lift Trucks (Guidance Note HS(G) 6)* HMSO, London

Health and Safety Executive (1988) *Approved Code of Practice: Rider-operated Lift Trucks – Operator Training,* HMSO, London

Health and Safety Executive (1992) *Provision and Use of Work Equipment Regulations 1992 and Guidance on Regulations* HMSO, London

Royal Society for the Prevention of Accidents (1975) *Training Manual for Power Truck Operators* RoSPA, Birmingham

Chapter 17
The Working Environment

Section 2 of the HASAWA places a duty on the employer to provide and maintain a working environment that is safe and without risks to health, including arrangements for the welfare of employees while at work. The Workplace (Health, Safety and Welfare) Regulations extend this duty further.

As such, the concept of the working environment covers a very broad range of aspects that have to be considered, in particular the location and layout of working areas, the use of colour, systems for waste disposal and traffic management, and the elimination or control of environmental stressors. Environmental stressors include extremes of temperature, lighting and ventilation, noise and vibration, and the presence of dusts and gases, all of which can have direct and indirect effects on the health of workers. The provision and maintenance of welfare amenity provisions are also important features of the working environment.

Location and layout of workplaces

In considering the possible location of a workplace, eg a factory, office block, warehouse, etc, consideration should be given to factors such as ease of access and exit for vehicles and pedestrians, the availability of public transport and vehicle parking areas, the potential for nuisance being created to the inhabitants of the neighbourhood, the vulnerability of the general public to major incidents, such as large fires or multiple road accidents, and the density of surrounding buildings.

The duties of the employer towards non-employees are clearly detailed in Section 4 of the HASAWA.

The layout of a workplace is significant in terms of preventing overcrowding, ensuring orderly manufacture of the product and the maintenance of satisfactory safety standards.

Regulation 10 of the Workplace Regulations stipulates that 'every room where persons work shall have sufficient floor area, height and unoccupied floor space for the purposes of health and safety'. In workplaces established before these Regulations came into force, ie 1 January 1993, the Schedule to the Regulations specifies a minimum of 11 cubic metres per person, with no space more than 4.2 metres from the floor being taken into account in any such calculation.

The use of colour in the workplace

Good choice of colour can promote a congenial environment which assists in achieving high standards of work performance. From the point of view of safety, the careful use of colour and colour schemes can do much to increase staff awareness of hazards and of the actions necessary should such hazards arise. Typical examples are the use of safety signs and symbols, as per the Safety Signs Regulations 1980, and in the yellow and black 'tiger striping' of floor and road surfaces to identify traffic hazards.

Careful choice of wall and ceiling colours also helps maintain good levels of illumination.

Waste disposal

Provision must be made for the collection, storage and disposal of waste materials from premises. While most liquid wastes are removable through the drainage system, solid wastes, particularly those which are flammable, can present a problem. The law requires that no waste material or refuse should be allowed to accumulate within a working area and that an adequate supply of waste containers should be provided. Refuse and waste materials should be removed from the premises on a daily basis and stored in a designated waste storage area. Where bulky wastes are produced, the use of an industrial waste compactor linked to a bulk waste container is recommended, rather than standard dustbins.

The local authority environmental health and engineering departments should be consulted wherever toxic or dangerous wastes must be removed and disposed of at an approved disposal site. In the majority of cases, specialist waste disposal contractors will provide such a service.

Environmental control

The following are the environmental elements that can be controlled.

Temperature
Exposure to extremes of temperature can result in heat stress and, at the other extreme, frostbite. Moreover, temperature control of the workplace, along with control over air movement and relative humidity, is important in the maintenance of comfort conditions. The temperature ranges in Table 5 are recommended according to the type of work undertaken.

Relative humidity
Relative humidity should be between 30 and 70 per cent. Below this range discomfort is produced due to the drying of the throat and nasal

Type of work	*Temperature range*
Sedentary and/or office work	19.4 to 22.8°C
Light work	15.5 to 20.0°C
Heavy work	12.8 to 15.6°C

Table 5 Comfort temperature ranges

passages, whereas above that range people will experience a feeling of stuffiness.

Lighting

Two aspects need consideration in lighting design:

(a) the quantity of light required for a given task, measured in lux (see following Note); and
(b) the quality of the lighting with regard to its distribution, the avoidance of 'glare' conditions, colour rendition and brightness.

Note: The standard unit of 'illuminance', that is the quantity of light required for a given task or area, is the lux. This equals one lumen per square metre. This unit replaced the foot candle which equated to the number of lumens per square foot. The term 'lumen' is the unit of luminous flux or light flow, describing the quantity of light received by a surface or emitted by a source of light.

Lighting standards are outlined in HSE Guidance Note HS(G) 38 'Lighting at work'. In this document a relationship is drawn between the average illuminance and the degree or extent of detail which needs to be seen in a particular situation or task. Average illuminance values and minimum measured illuminance values, both measured in lux (lx), are shown in Table 6. The minimum measured illuminance is the lowest illuminance permitted in the work area, taking the requirements of health and safety into consideration.

Attention must also be paid to the relationship between the lighting of the work area and adjacent areas, perhaps used for storage. Substantial differences in illuminance levels between such areas may produce visual discomfort and even affect safety levels where there is frequent movement of pedestrians and vehicles, such as fork-lift trucks. To reduce risks to both operators and vehicles, as well as possible visual discomfort, maximum illuminance ratios are therefore recommended in the Guidance Note. (See Table 7.)

Where there is conflict between the recommended average illuminances shown in Table 6 and the maximum ratios of illuminance in

General activity	*Typical locations/ types of work*	*Average illuminance (lx)*	*Minimum measured illuminance (lx)*
Movement of people, machines and vehicles (1)	Lorry parks, corridors, circulation routes	20	5
Movement of people, machines and vehicles in hazardous areas; rough work not requiring perception of detail	Construction site clearance, excavation and soil work, loading bays, bottling and canning plants	50	20
Work requiring limited perception of detail (2)	Kitchens, factories assembling large components, potteries	100	50
Work requiring perception of detail	Offices, sheet metal work, bookbinding	200	100
Work requiring perception of fine detail	Drawing offices, factories assembling electronic components, textile production	500	200

Table 6   Average illuminances and minimum measured illuminances
(*Source* HSE Guidance Note HS(G) 38 'Lighting at work')

Notes
1. Only safety has been considered, because no perception of detail is needed and visual fatigue is unlikely. However, where it is necessary to see detail to recognise a hazard or where error in performing the task could put someone else at risk, for safety purposes as well as to avoid visual fatigue, the figure should be increased to that for work requiring the perception of detail.
2. The purpose is to avoid visual fatigue: the illuminances will be adequate for safety purposes.

Table 7, the higher value should be taken as the appropriate average illuminance.

'Glare' is a problem which is occasionally encountered in lighting installations, and is the effect of light which causes discomfort or impaired vision. It is experienced when parts of the visual field are excessively bright compared with the general surroundings. It frequently occurs when the light source is directly in line with the task being undertaken or when light is reflected off a given object or surface. Glare can occur in the following three forms.

Situations to which recommendation applies	Typical location	Maximum ratio of illuminances		
		Working area		Adjacent area
Where each task is individually lit and the area around the task is lit to a lower illuminance	Local lighting in an office	5	:	1
Where two working areas are adjacent, but one is lit to a lower illuminance than the other	Localised lighting in a works store	5	:	1
Where two working areas are lit to different illuminances and are separated by a barrier but there is frequent movement between them	A storage area inside a factory and a loading bay outside	10	:	1

Table 7 Maximum ratios of illuminance for adjacent areas
(*Source* HSE Guidance Note HS(G) 38 'Lighting at work')

1. Disability glare
This is the visually disabling effect caused by bright bare lamps directly in the line of sight. The impaired vision (dazzle) which results from this effect may be hazardous in the driving situation, when working in high-risk operations or at heights.

2. Reflected glare
This is the reflection of bright light sources on wet or shiny work surfaces, such as plated metal or glass. The effect is to totally conceal the detail in or behind the object which is glinting. Wherever possible, light sources of low brightness levels should be used and the geometry of the lighting installation should be arranged so that there is no glint at the viewing position.

3. Discomfort glare
This is caused by too much contrast of brightness between an object and its background. The phenomenon is generally associated with poor lighting design. Visual discomfort can result but the ability to see detail may not be impaired. Over a period of time operators exposed to discomfort glare can experience eye strain (visual fatigue), general fatigue and headaches.

The situation can be resolved in an existing work situation by:

(a) keeping luminaires as high as is practicable;
(b) maintaining luminaires parallel to the main direction of lighting; and
(c) careful design of shades which screen the lamp.

Reference should be made to the Illuminating Engineering Society's (IES) schedule of Limiting Glare Indices in the design or improvement of lighting installations. These are indices representing the degree of discomfort glare which will be just tolerable in the process or location under review. Where these indices are exceeded, occupants may experience visual fatigue and/or headaches.

Lighting distribution is another important feature of lighting design. Distribution is concerned with the way light is spread, and is classified under the British Zonal Method from BZ1 (all light downward in a vertical column) to BZ10 (light in all directions). For good general lighting, regularly spaced luminaires should be used to give an evenly distributed illuminance. This evenness of illuminance depends upon the ratio between the height of the luminaire above the working position and the spacing of fittings. The IES spacing:height ratio provides a general guide to such arrangements, the normal ratio being $1\frac{1}{2}$:1 or 1:1, depending upon the type of luminaire.

Brightness (luminosity) is a subjective sensation and cannot be measured. However, a brightness ratio can be considered, which is the ratio of apparent luminosity between a task object and its surroundings. All surfaces have a particular level of reflectance, ie the ability of a surface to reflect light. Given a task illuminance factor of 1, the comparable reflectance values should be as shown in Table 8.

Colour rendition refers to the appearance of an object under a given light source, compared to its colour under a reference illuminant, eg natural light, and enables the true colour to be correctly perceived.

Generally, the colour rendering properties of fitments should not be in stark contrast to those of natural light. They should also be as effective at night as during the day, as then there will be no daylight contribution to total workplace illumination.

Ceilings	0.6
Walls	0.3 to 0.8
Floors	0.2 to 0.3

Table 8   Effective reflectance values (Illuminating Engineering Society)

Ventilation

Two aspects must be considered to ensure satisfactory ventilation of the workplace, namely the provision of sufficient air for people to breathe and work in reasonable comfort (comfort ventilation) and, secondly, means for the removal of air which may have become contaminated as a result of processes (exhaust/extract ventilation). The first requirement is dealt with in the Workplace Regulations, whereas the requirement to provide effective exhaust ventilation is covered by the COSHH Regulations.

In the case of comfort ventilation, Regulation 6 of the Workplace Regulations states that:

1. Effective and suitable provision shall be made to ensure that every enclosed workplace is ventilated by a sufficient quantity of fresh or purified air.
2. Any plant used for the purpose of complying with the above paragraph shall include an effective device to give visible or audible warning of any failure of the plant where necessary for reasons of health or safety.

The requirements, therefore, are clear. There must be sufficient circulation of fresh air to maintain comfort conditions and a healthy working environment. Where there is emission of dust, fumes, gases and other forms of airborne pollution from processes then, additionally, effective systems of exhaust ventilation must be provided and maintained.

Mechanical ventilation systems

Reliance on natural ventilation, ie from doors, windows and other openings in the fabric of a workplace, is totally unsatisfactory where there may be emissions of dust and fumes from processes. In these cases mechanical ventilation systems must be operated. Such systems take two principal forms – extract (exhaust) ventilation and dilution ventilation.

Extract ventilation (exhaust ventilation): There are three main types of extract ventilation system – receptor systems, captor systems and low- volume high-velocity (LVHV) systems (see Figure 28).

1. *Receptor systems:* With this type of system the contaminant enters the system without inducement. The fan maintains a flow of air to transport the contaminant from its point of emission through a hood and ducting to a collection system. In some cases the hood actually forms a total enclosure around the source, eg a laboratory fume cupboard, or a partial enclosure, eg a paint spray booth.
2. *Captor systems:* As the term implies, moving air captures the

RECEPTOR SYSTEMS

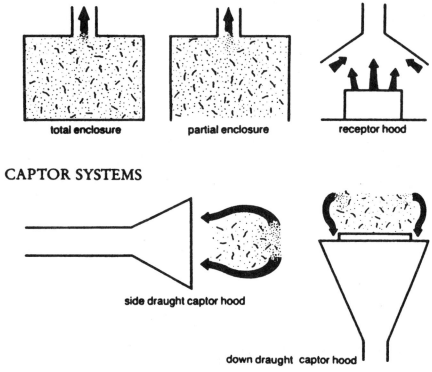

CAPTOR SYSTEMS

LOW-VOLUME HIGH-VELOCITY SYSTEMS

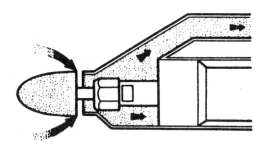

Figure 28   Extract ventilation systems

contaminant at some point outside the receiving hood and induces its flow into same. The rate of air flow into the hood, which can take a number of different shapes, must be high enough to capture the contaminant at the furthest point of origin. Moreover, the air velocity induced at this point must be sufficient to prevent the contaminant flowing in another direction away from the hood. Captor systems are frequently used for welding activities, other than in a specifically designed welding booth or bay, with certain types of grinding machines and in soldering operations, the principal objective being to take the offending airborne contaminant away from the operator's breathing zone.

3. *Low-volume high-velocity (LVHV) systems:* With certain high-speed grinding machines or pneumatically operated chipping tools very high capture velocities are required. To prevent environmental contamination and the risk of dust being inhaled by the operator, LVHV systems are used. These systems operate on the principle of extracting the dust through small apertures very close to the point of emission, thereby achieving high velocities at the source and with low air flow rates.

Such systems need regular maintenance, together with comprehensive operator training.

Dilution ventilation
Dilution ventilation implies diluting the concentration of an offending agent, perhaps in gaseous form, with large quantities of fresh air to a point where the gas or vapour is no longer dangerous. This system is most successfully used to control vapours from low toxicity solvents in small quantities which are uniformly evolved from a process. Dilution ventilation is not recommended as a means of controlling dust and fume emission.

Ventilation and the COSHH Regulations
Under the COSHH Regulations an employer must either prevent or control exposure to hazardous substances. He must also ensure the controls are properly used, and maintain, examine and test these control measures. Local exhaust ventilation (LEV) systems ie receptor systems and captor systems, are the most common form of control for airborne contaminants.

Examination and testing may be by means of visual checks, inspections, testing and servicing with a view to ensuring the system is maintained in effective working order. Manufacturers and suppliers of LEV systems should stipulate the method of testing and, in many cases, may undertake such testing on behalf of employers. In addition to effective preventive maintenance, there are statutory requirements to undertake formal examination and testing of LEV systems in certain industries, eg ceramics. Records of examination and tests should be kept, and must be available for inspection by employees, Employment Medical Advisers and inspectors.

Noise

Noise is generally defined as 'unwanted sound'. Sources of noise in the workplace include:

(a) noise taking a structure-borne pathway;
(b) noise produced as a result of vibration in machines;
(c) radiation of structural vibration into the air;
(d) turbulence created by air or gas flow;
(e) noise taking an airborne pathway; and
(f) noise produced by vibratory hand tools, such as hand-held grinders and chain saws.

Protection strategies

People who are exposed to noise in excess of 90 dBA (see below) on a continuous or intermittent basis stand the risk of going deaf, ie noise-induced hearing loss or occupational deafness. (The decibel (dB) is a unit of sound pressure. Sound pressure level is a measurement of the magnitude of the air pressure variations or fluctuations which make up a particular sound. dB(A) implies measurement of sound pressure level using the 'A' network of a sound pressure level meter. The 'A' network of such an instrument is the one which most closely follows the performance of the human ear.)

Protection strategies directed at preventing workers from sustaining noise-induced hearing loss are, in order of importance, reduction of the noise at source, ie designing quieter machines, isolation of the noise source, ie by the use of soundproof enclosures and close shields, provision and use of hearing protection for workers exposed to noise, and reduction of the time during which people are exposed to noise.

It should be appreciated that the decibel scale is a logarithmic scale (not a linear scale) and, on this basis, each increase in pressure of three dBA represents a doubling of the sound intensity. Therefore, to ensure that operators do not receive more than a 90 dB dose of noise per eight- hour working day, sound pressure level should be related to the duration of exposure as shown in Table 9.

90 dB(A) – maximum duration of dose –	8 hours
93 dB(A) – maximum duration of dose –	4 hours
96 dB(A) – maximum duration of dose –	2 hours
99 dB(A) – maximum duration of dose –	1 hour
102 dB(A) – maximum duration of dose –	30 minutes
105 dB(A) – maximum duration of dose –	15 minutes

Table 9 Comparable noise doses

Sources and pathways	Control measures
Vibration produced as a result of machinery operation	Reduction at source, eg substitution of nylon components for metal
Structure-borne noise (vibration)	Vibration isolation, eg resilient mounts and connections
Radiation of structural vibration	Vibration damping to prevent resonance
Turbulence created by air or gas flow	Use of silencers, similar to vehicle silencers
Airborne noise pathway	Noise insulation – heavy barriers Noise absorption – porous lightweight barriers

Table 10 Noise sources, pathways and control measures
Source Department of Safety & Hygiene, University of Aston in Birmingham

Noise control
Noise control implies, first, a consideration of the source of the noise (in machinery there may be many sources of noise) and, secondly, of the pathway taken by the noise to the recipient. The use of personal protection – ear muffs, acoustic wool, ear plugs – relies heavily on the exposed person using this equipment while exposed to the particular noise level and, as such, will never be a perfect solution to the problem.

Noise control strategies should take into account the sources, pathways and control measures indicated in Table 10.

Noise at Work Regulations 1989

These Regulations impose specific requirements on employers to:

(a) make and regularly review noise assessments;
(b) keep records of noise assessments and any reviews thereof;
(c) generally reduce the risk of damage to the hearing of their employees from exposure to noise;
(d) reduce exposure of their employees to noise, other than by the use of ear protectors;
(e) provide employees with personal ear protectors;
(f) mark designated ear protection zones;
(g) use and maintain equipment provided to protect hearing; and
(h) provide information, instruction and training for employees.

Employees have a duty to use the equipment, including ear protectors, and to report defects discovered in such equipment to the employer.

Machinery manufacturers and suppliers must inform the employer of noise levels likely to be generated by their machinery and equipment.

Noise action levels
Three 'action levels' are specified in the Regulations thus:

(a) the first action level – means a daily personal noise exposure of 85 dB(A);
(b) the second action level – means a daily personal noise exposure of 90 dB(A);
(c) the peak action level – means a level of peak sound pressure of 200 pascals.

Noise exposure assessment
Regulation 4 requires that every employer shall, when any of his employees is likely to be exposed to the first action level or above, or to the peak action level or above, ensure that a competent person makes a noise assessment which is adequate for the purposes:

(a) of identifying which of his employees are so exposed; and
(b) of providing him with such information with regard to the noise to which those employees may be exposed as will facilitate compliance with his duties under the Regulations.

Welfare amenity provisions
Amenities provided for workers include arrangements for sanitation (toilets, urinals), washing facilities and showers, the provision of drinking water and meals, and the storage and drying of clothing. Specific requirements are detailed in the Workplace Regulations (Regulations 21 to 25) and ACOP. Such requirements cover the number of water closets, urinals and wash basins on the basis of the number of people employed of both sexes, provisions relating to the accessibility of same, lighting and ventilation of amenity areas, and maintenance requirements. In certain industries controlled by specific Regulations, eg Control of Lead at Work Regulations 1980 and Construction (Health and Welfare) Regulations 1966, more specific provisions are laid down.

The principal Acts also lay down certain requirements relating to the provision of seats where it is reasonable for a substantial part of the work to be carried out in a sitting position and, in the case of offices and shops, the provision of suitable and sufficient facilities for persons who may take meals on the premises.

Summary

1. There is a legal duty on the employer under the HASAWA to provide and maintain a working environment that is safe and without risks to health, including arrangements for the welfare of employees while at work.

2. Control of the working environment should take into account the location and layout of the workplace, including means of access and exit, the prevention of overcrowding, the use of colour, systems for waste disposal and the need for sound traffic-management procedures.

3. Environmental stress can result in poor levels of operator performance, accidents and occupational ill health.

4. To prevent environmental stress, attention must be paid to systems for temperature and humidity control, lighting arrangements, ventilation, control over noise and vibration, and the removal of dangerous airborne contaminants.

5. Good standards of welfare amenity provision are essential features of the working environment.

References

Bilsom International (undated) *In Defence of Hearing* Bilsom International Ltd, Henley-on-Thames

Bruel & Kjaer (1984) *Measuring Sound* Bruel & Kjaer, Naerum, Denmark

Burns, W (1973) *Noise and Man* John Murray, London

Electricity Council *Better Office Lighting* Electricity Council, London

Health and Safety Commission (1992) *Workplace (Health, Safety and Welfare) Regulations 1992 and Approved Code of Practice* HMSO, London

Health and Safety Executive (undated) *Ventilation of the Workplace (Guidance Note EH/22)* HMSO, London

Health and Safety Executive (1975) *Principles of Local Exhaust Ventilation* HMSO, London

Health and Safety Executive (1987) *Lighting at Work, (Guidance Note HS(G) 38)* HMSO, London

Health and Safety Executive (1989) *Noise at Work Regulations 1989 and various Noise Guides* HMSO, London

Health and Safety Executive (1991) *Noise at work: advice for employees* HSE Information Centre, Sheffield

Health and Safety Executive (1992) *Listen up* HSE Information Centre, Sheffield

Lyons, S (1984) *Management Guide to Modern Industrial Lighting*, Butterworths, Sevenoaks

Noise at Work Regulations 1989, HMSO, London

Waring, R A (1970) *Handbook of Noise and Vibration Control* Trade & Technical Press, London

Chapter 18
Safety in Offices, Workshops and in Catering Operations

Most people would consider the average office to be a reasonably safe workplace compared, say, with a construction site or foundry. However, while fatal accidents are uncommon, minor accidents caused by trips and falls, the unsafe use of electricity, obstructed passages and stairways, failure to use access equipment and poor housekeeping are common.

Fire is, by far, the greatest hazard in offices. Because of the design, location, construction, layout and age of many offices, a fire can spread rapidly from floor to floor. The need for well-developed fire and emergency procedures in offices cannot be over-emphasised. This is particularly appropriate in large multi-occupied offices housed in old converted buildings.

Safety requirements for offices

The following areas should be paid attention to regarding safety in offices.

Housekeeping

One of the greatest causes of accidents in offices, particularly falls, is that of bad housekeeping. The term 'housekeeping' implies 'everything in the correct place and a place for everything'. Bad housekeeping is frequently the cause of office fires which can be caused by general untidiness, poor storage of flammable wastes, such as waste paper, and people smoking. The following housekeeping rules should be applied in offices:

(a) Keep the work area tidy. Items that are not in use should be stored away.
(b) Waste should be stored in waste containers, which should be emptied on a regular daily basis.
(c) Heavy items, such as ledgers, should not be placed on the top of cabinets or cupboards as they could fall on to someone using the cupboard.
(d) Harmful items, such as broken crockery, light bulbs and milk bottles, should be separately stored and disposed of.

(e) Passages, stairways, entrances and exits, in particular emergency exits, should be kept clear and free from surplus stationery, sacks of office refuse, surplus office equipment and furniture, and other large items.
(f) Spillages should be cleared up immediately.
(g) Damaged floor coverings, such as carpets, should be replaced immediately due to the tripping or slipping hazard created.
(h) An established cleaning schedule should be maintained.

Furniture and fittings

Bad office planning and layout with regard to the siting of furniture and fittings can result in numerous minor accidents. The following points should be considered with regard to office layout and the location of items such as filing cabinets, desks and equipment:

(a) Furniture should be arranged so that employees can move freely within the office and from one office to another.
(b) Doors and drawers, particularly to filing cabinets, should be kept closed when not in use.
(c) When using a filing cabinet, only one drawer should be opened at a time due to the risk of the cabinet tipping forward, particularly when the two top drawers are open at the same time.
(d) Filing cabinets should not be overloaded. Heavy items should be stored in the bottom drawer.
(e) The bottom drawer of a filing cabinet should always be closed immediately to avoid the risk of people tripping over it.
(f) Damaged and broken furniture and fittings, such as shelves, chairs and filing cabinets, should be repaired or replaced immediately.

Electrical appliances

Misuse and abuse of electricity in offices is one of the most significant causes of office fires. There should be a total prohibition on staff undertaking electrical repairs or modifications. The increased use of electrical appliances in offices has also contributed to the risk of electrical overloading. The following rules should be observed:

(a) Only trained and competent staff should attempt to repair electrically operated machinery.
(b) Machines should be switched off from the mains when left unattended for long periods.
(c) Cables should be so positioned that they do not trip people up. Where this is not possible, suitable permanent cable covers should be installed. Flexes should be shortened so that they do not trail across the floor or under desks.

(d) The use of freestanding radiant-type electric fires, sometimes used to supplement central heating in the winter months, should be prohibited.

(e) Cables to electrical appliances should be maintained in a sound condition and replaced when they become frayed or damaged.

(f) The use of multi-point adaptors should be either prohibited or carefully controlled in order to avoid overloading sockets.

(g) Electrical appliances should be examined every six months and a record of such examinations maintained.

Lifting and carrying

A substantial number of permanent back injuries are sustained by office staff lifting and carrying heavy items, such as electric typewriters, stationery packages and office furniture. To prevent the risk of such injuries, adequate manual-handling equipment, such as trolleys and sack trucks, should be provided. The general rule must be that no one should lift anything which is likely to cause injuries to the back, hands, arms, legs or feet.

Dangerous substances

A variety of products used in offices on a daily basis can be potentially dangerous. Such products include cleaning fluids, adhesives, quick drying inks and correcting fluids, all of which emit powerful fumes or vapours. In certain circumstances, such substances can be highly flammable. Other flammable substances include floor polishes and some aerosol-based cleaning compounds. To reduce the risks associated with dangerous substances, the following precautions are necessary:

(a) Staff should always read the manufacturer's instructions prior to using a potentially dangerous product.

(b) In certain situations it may be necessary to wear personal protective equipment, such as gloves, apron and goggles, when dealing with substances.

(c) Staff should be aware of the hazard warning symbols shown on the packages for potentially dangerous substances, eg flammable, toxic symbols.

(d) Waste should be disposed of safely. This particularly applies to cloths soaked in solvent-based cleaning fluids. In this case, such items should be stored in a metal container with a close-fitting lid and disposed of on a daily basis.

(e) Dangerous substances should be handled in a well-ventilated area.

(f) Any ill-effects experienced by staff following the use of substances should be reported immediately.

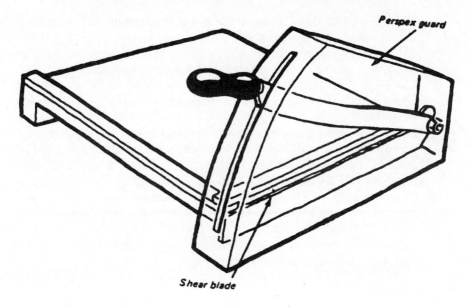

Figure 29   Guarding to a hand-operated guillotine

It should be appreciated that the COSHH Regulations apply to offices and to hazardous substances used in offices.

Office equipment
Certain items of office equipment, in particular hand-operated guillotines, can inflict serious injury. Guillotines should be effectively guarded (see Figure 29) and staff trained in their safe use.

Other items, such as scissors, letter openers and knives, should only be used for their main purpose and certainly not, for example, as screwdrivers or for opening tins.

Fire precautions
The majority of fires in offices occur outside normal working hours. The initial cause of the fire will, in many cases, have been created during office hours. To minimise the risk of fire, the following procedures and practices, some of which have already been mentioned earlier in this chapter, should be followed:

(a) On no account should fire exits and designated escape routes be obstructed.
(b) All flammable wastes should be carefully controlled in terms of storage and removal.
(c) The use of radiant-type electric fires should be prohibited.
(d) Smoking in the office should be carefully controlled, including

the use of ash trays. Persons who are careless in their smoking habits should be disciplined.

(e) All electrical equipment should be disconnected from the socket when not in use or left unattended for long periods.

(f) Clothing and other items should not be dried close to a direct heat source. In particular, they should not be placed over space heaters.

(g) Prior to locking the office at the termination of work, a trained person should undertake an inspection to ensure that no fire hazards have been created. In larger offices, designated fire wardens should carry out this task.

(h) Flammable items should be kept to a minimum for general use. A metal storage cupboard should be provided for storage.

(i) All staff should be aware of the nearest fire alarm point, the nearest fire appliance and the emergency evacuation plan for the building, including assembly points.

Personal conduct

It is regrettable that many office workers do not envisage their work-place as being potentially dangerous compared, say, with a factory. This attitude can result in accidents, and staff should be trained in the following basic aspects of personal conduct at the induction training stage:

(a) Dress sensibly for work. Do not wear items that may catch in office machinery and do not wear unsuitable footwear.

(b) Look where you are going! Do not read while walking or carry items at such a level that you cannot see where you are going.

(c) Do not run or turn corners quickly when you cannot see what is round the corner.

(d) Use the access equipment provided when storing items normally out of reach, and not revolving typist chairs or existing shelves.

(e) Open doors carefully! There may be someone standing on the other side.

Office health and safety audit

To ensure maximum standards of health and safety in the office, the audit shown below should be used.

1. Have the premises been registered under the Offices, Shops and Railway Premises Act 1963?

2. If more than 20 persons, or more than 10 persons on any floor other than the ground floor, are working in the building, has a Fire Certificate been obtained?

3. Has the HSE poster *Health and Safety Law – What you should know* been displayed prominently in the office, in accordance with the Health and Safety (Information to Employees) Regulations 1989? Have leaflets been provided to staff?
4. Are all furniture, fittings and furnishings in a clean state?
5. Is there sufficient space for people to work, bearing in mind the overcrowding standards quoted in the Workplace Regulations?
6. Is the temperature reasonable?
7. Is the means of heating safe and without risks to health?
8. Is a thermometer provided in a conspicuous place on each working floor of the premises?
9. Is ventilation adequate?
10. Is the lighting adequate and in accordance with standards established by the HSE Guidance Note 'Lighting at work'.
11. Are enough toilets provided for both male and female staff?
12. Are there suitable and sufficient washing facilities for both male and female staff, together with soap and drying facilities?
13. Is there an adequate supply of wholesome drinking water?
14. Is there suitable accommodation for external clothing not worn in the office?
15. Are there adequate and suitable seating arrangements?
16. Are all floors, passageways and staircases in a good state of repair and free from obstruction?
17. Do staircases have handrails?
18. Have dangerous parts of machinery been guarded?
19. Is there a suitable number of fully equipped first aid boxes?
20. Are there enough trained first-aiders available during normal working hours?
21. Is the Emergency Evacuation Procedure established and displayed?
22. Are the roll call lists up to date?
23. Are fire precautions records being maintained?
24. Are all fire appliances correctly located and regularly serviced?
25. Are all areas free from accumulations of combustible materials?
26. Are all fire exit doors clearly marked, unobstructed and operational?
27. Can all escape route doors be opened easily?
28. Are all fire-resisting or smoke-stop doors closed and fitted with appropriate door closing gear?
29. Can the fire alarm be heard in all parts of the premises?
30. Has the fire alarm been tested recently?
31. Has a fire drill been carried out in the last 12 months?
32. Have persons been trained in the correct use of fire appliances?

Workshop safety

Many hazards can occur in workshops arising from congestion, inadequate storage, fire, machinery and structural risks, poor environmental conditions and the design and layout of the workshop.

Factors for consideration in assessing the safety of workshops are indicated below.

Structural features

1. **Floors** – clean, sound finish; adequate floor drainage where necessary.
2. **Inspection pits** – safe access; clean; intrinsic flameproof lighting; pit covers/boards or secure fencing when not in use; drainage with sump or pump.
3. **Elevated storage areas** – safe access; adequate lighting; safety rails (minimum 1 metre high) with intermediate rails and toe boards.
4. **Fixed ladders** – sound; back rings fitted from 2 metres upwards; adequate lighting.
5. **Portable ladders** – sound styles and rung-and-style connections; inspection system for ladders; correct storage.
6. **External areas** – lighting; drainage; yard surfacing; marking out; directional signs; traffic hazards; segregation of pedestrian and traffic routes; parking arrangements; storage of waste.

Environmental aspects

1. **Temperature control** – adequate; minimum 16°C after first hour of working; wall-mounted thermometer; fumes from heating appliances; use of portable heating appliances.
2. **Lighting** – suitable and sufficient; specific lighting arrangements, eg work benches; windows and roof lights clean.
3. **Ventilation** – adequate; LEV systems installed eg welding operations.
4. **Noise** – machinery and appliances; physical separation; noise reduction; ear protection provided and worn; notices displayed.

Machinery and equipment

1. **Abrasive wheels** – correctly mounted; maximum spindle speed notice displayed (variable speed grinders); tool rest/wheel – maximum 0.125 inches clear; motor isolation when changing wheel; fixed guard; eye protection provided and used; mounting instructions displayed; adequate bench and specific lighting.
2. **Lifting tackle** – SWL marked; inspection and certification every 6 months; stored off floor; slinging table displayed.

3. **Drills and drilling machines** – flexes and plugs; chuck and spindle guard.
4. **Gas welding** – welding screen/curtain; cylinder racks/chains; full face protection (tinted); cylinder carrier/cradle; hoses; correct cylinder storage.
5. **Electric welding** – welding screens/curtain; cables and clips; equipment and workpiece earth bonded; full face protection (tinted); gloves and apron.
6. **Air compressors** – belt drive guarded; isolation to motor; SWP marked on reservoir; calibrated safety valve; regular draining of reservoir; examination and certification procedure.
7. **Compressed air equipment** – eye protection provided and worn; abuse/misuse; metering valve and tyre pressure gauge; storage of compressed air tools; tyre inflation cage.
8. **Jacking operations** – saddles clean and sound; examined monthly; correctly stored; hydraulic jacks – examination and certification.
9. **Woodworking machinery** – circular saws – guarding below saw table, crown and adjustable guard fitted; use of push sticks; stopping device.
10. **Lathes** – chuck and face plate guard; clamp; eye protection; swarf removal.
11. **Guillotines** – sound construction; fixed or adjustable guard.
12. **Vehicle lifts** – sound construction; wheel stops fitted and operational; chocks provided; warning device/alarm during operation.
13. **Valve grinders** – sound construction; eye protection.
14. **Ramps** – sound construction; wheel stops fitted and used.
15. **Vehicle washing** – drainage; winter icing arrangements; chemical storage and dosing.

Hand tools
1. **Files** – condition; chips and cracks; handles in sound condition; training in correct use.
2. **Hacksaws** – condition; state of blade; training in correct use.

Electrical safety
1. **Hand tools** – low voltage (110v 24v); efficient switches; earthing; double insulation; flexes sound; connections; socket overloading.
2. **Hand lamps** – bulb cage; earthed/low voltage; overhead drop leads; flexes and connections sound.
3. **Battery charging** – flexes and connections sound; earth clamp; adequate ventilation of charging area.
4. **Machine controls** – mushroom headed STOP buttons; inset/shrouded START buttons; clearly identified; isolation.
5. **Control boxes** – kept shut; ON-OFF switch clearly marked.

Fire prevention and control

1. **Flammable materials** – identified; separate storage; potentially flammable atmospheres; spillage control; notice displayed.
2. **Welding equipment** – sitting away from combustible/flammable substances; purging of pits (compressed air).
3. **Flammable wastes** – storage and disposal; incineration hazards.
4. **Diesel store** – tank sound; pump/dispenser sound.
5. **Fire protection and appliances** – fire appliances wall mounted, serviced annually; training in correct use of appliances; fire exits and escape routes correctly marked; key in box/crash bar; exits checked regularly; fire alarm tested regularly; fire drills annually at least.

Hazardous substances

1. **Cleaning and degreasing agents** – dermatitis risks; use of gloves, eye protection, apron; emergency eye/face wash facility; storage tank sound; no smoking; adequate ventilation; labelling of containers and transfer containers.
2. **Solvents** – general awareness of relative flammability and toxicity; storage in accurately labelled containers.
3. **Anti-freeze** – general awareness of relative flammability; storage in accurately labelled containers.
4. **General storage** – ventilated; floor drainage where appropriate; spillage control; personal protection requirements; limited access.

Storage areas

1. **Racking** – stable – fixed to floor/wall; configuration; evidence of overloading; spring racks secure.
2. **Separation** – flammable/dangerous substances stored separately; metal cabinets for small quantities.

Amenity area

1. **Sanitation** – clean; adequate facilities; decoration.
2. **Washing facilities** – clean; hot and cold water; soap, nailbrushes, drying facility; barrier cream in dispenser.
3. **Clothing storage** – adequate; drying facilities for clothing; secure lockers; separation of soiled protective clothing from external clothing where appropriate.
4. **Drinking water** – adequate facilities; fountains.
5. **Mess room/rest room** – adequate; seats provided; ventilated and heated; cooking/food heating facility; sink with hot and cold water; utensil drying facility.
6. **First aid** – first aid box provided and maintained; first-aiders sufficient and trained.

Administration

1. **Registers etc** – planned maintenance record; hoists and slings; abrasive wheels; ladders; power washer maintenance record; high pressure pumps service records; accident book.
2. **Statement of Health and Safety Policy** – clearly displayed.
3. **Abstracts** – 'Health and Safety Law – What you should know' poster displayed.

Catering safety

Catering activities can be dangerous, particularly where kitchens may be badly designed and staff untrained in the hazards and precautions necessary.

Catering injuries

Typical injuries associated with catering activities include:

(a) scalds to the hands, forearms, feet, legs and trunk through contact with boiling water, hot fats and liquids;
(b) cuts to the hands through the use of knives, slicing machinery and can-opening operations, together with those caused through contact with broken glass and crockery;
(c) burns to the hands and forearms from ovens, hotplates, oven-ware, plates and hot liquids;
(d) bruising, abrasions and fractures due to slips, trips and falls on wet and greasy floors; and
(e) back injuries through incorrect manual handling techniques.

The following types of accidents and injuries to catering staff are common:

Falls on stairs	23%
Falls on kitchen floors	20%
Cuts and hand injuries from slicing and mixing machines	19%
Burns and scalds from various sources	11%
Back injuries from lifting heavy packs	11%
Hand injuries from the use of knives	8%
Falls from step ladders	5%
Strains, sprains and back injuries from moving furniture and equipment	3%

The precautions necessary

1. Floors

Falls on kitchen floors can result in broken limbs, head injuries, strains and sprains. It is, therefore, essential that catering staff wear sensible low-heeled shoes with non-slip soles. The floor should be adequately drained, and cleaning and housekeeping procedures should ensure that spillages of fats and liquids are cleared directly after they occur.

2. **Hand tools**
And what about those nasty cuts from boning knives, meat cleavers and saws used in the preparation of meat dishes? Even sharpening a knife using a butcher's steel can be dangerous, particularly if the steel does not incorporate a guard between the handle and the steel. Staff training in the correct use of knives and sharpening procedures is essential.

3. **Machinery**
Catering machinery can include slicing machines, mechanical potato chippers, bowl mixers, waste disposal units and dish washing machines. All these items incorporate moving parts which should be effectively guarded or, alternatively, fitted with appropriate safety devices which prevent the hands coming into contact with moving parts in compliance with the PUWER 1992. Over one-third of machinery-related accidents are associated with slicing machines and are caused by unsafe operation and cleaning of these machines. Causes include inadequate guarding of the machine, removal of the guards while the machine is operating, use of the hand to push forward the meat being sliced, unsafe methods of cleaning the blade, and lack of training in the safe use of such machinery.

4. **Environmental factors**
A significant indirect cause of accidents in kitchens is frequently the poor levels of working environment provided. Temperatures in kitchens frequently reach 90°F 32°C, many are inadequately ventilated and the high levels of humidity from steaming appliances and boiling liquids can result in staff becoming lethargic and careless. Controlled ventilation giving 12 to 20 air changes per hour, particularly during summer months, together with exhaust ventilation over ranges is essential. Well-designed lighting systems in preparation, storage and external areas should be provided and maintained.

5. **Fire**
The risk of fire in catering activities needs urgent consideration due to the very nature of much of the equipment used – ovens, frying equipment, grills and other forms of open-flame appliance. Many food ingredients and cooking aids are highly flammable, such as cooking oil, butter, margarine and lard. They spread rapidly on a floor taking a fire to other parts of a kitchen. In particular, staff should be trained to deal with small fat fires on cooking ranges, there should be an adequate number of fire appliances and fire blankets available, fire exits should be

marked and kept clear, and regular fire drills held. It is vital that staff should know the right type of appliance to use with different classes of fire.

6. Manual handling

Catering staff are frequently required to handle heavy weights, such as meat joints, turkeys, packs of dry goods, sacks of vegetables and heavy utensils. A substantial number of injuries are caused by incorrect manual handling or through staff attempting to lift items which are too heavy. With any manual handling operation, the rule must be 'If you can't lift it, get help.'

One of the principal causes of back injury is lifting heavy loads from the bottom of a chest freezer where it is virtually impossible to practise the principles of safe lifting. Wherever possible, heavy frozen items, such as meat joints, turkeys and bagged vegetables should be stored in purpose-built cold stores or upright freezer units to eliminate the risk of back injury.

7. Personal protective clothing

Many accidents to catering staff are caused through the incorrect use of personal protective clothing – long sleeves, open coats, inadequate hair covering. For both safety and hygiene reasons, protective clothing should be reasonably tight-fitting and the hair completely covered by a net. This includes male catering staff!

8. Jewellery

Significant injuries to fingers are associated with the use of rings. Only a plain wedding ring should be worn.

9. Hazardous substances

A wide range of hazardous substances are encountered in catering operations and, in some cases, staff are inadequately trained and super-vised in their correct use. Detergents and detergent sanitisers may be strongly acidic or alkaline. Used at an incorrect dilution, they can cause skin damage in the form of rashes to the hands and arms and, in some cases, dermatitis. Dilution instructions must, therefore, be carefully followed. Staff engaged in the use of hazardous cleaning substances should be provided with elbow-length gloves and be prevailed upon to use them whenever there is a risk of hand contact with such substances. In some cases it may be necessary to undertake health risk assessments of substances hazardous to health under the COSHH Regulations 1994.

Summary

1. Most people consider office work to be relatively safe and do not appreciate the risks, particularly that of fire.

2. Poor standards of housekeeping is one of the principal causes of accidents in offices.

3. Many items of office furniture and fittings can be dangerous if not properly used.

4. Abuse and misuse of electrical appliances and equipment can result in fires and the risk of electrocution.

5. All staff should be trained in correct manual-handling techniques, and handling equipment should always be provided to reduce the risk of handling accidents.

6. Staff should be aware of the dangerous substances in use and the precautions necessary.

7. A clearly established fire procedure is essential.

8. Unsafe behaviour by office staff should not be tolerated, with disciplinary action being taken against offenders.

9. An office safety audit should be undertaken at six-monthly intervals by a responsible person.

10. The principal hazards in workshops are associated with congestion, poor housekeeping, inadequate lighting, various types of machinery and equipment, storage of flammable substances and electricity.

11. Workshops are subject to the requirements of the Workplace (Health, Safety and Welfare) Regulations 1992 and the Provision and Use of Work Equipment Regulations 1992.

12. Storage areas should be fitted with suitable racking which should be well maintained.

13. The principal types of accident in catering operations are falls, cuts and hand injuries.

14. Kitchen floors can be extremely dangerous if poorly maintained and not cleaned regularly.

15. A wide range of hazardous substances is encountered in catering operations, in various forms of detergent.

References

Health and Safety Commission (1992) *Workplace (Health, Safety and Welfare Regulations 1992 and Approved Code of Practice* HMSO, London

Health and Safety Executive (1987) *Health and Safety in Kitchens and Food Preparation Areas* HMSO, London

Health and Safety Executive (1987) *Catering Safety: Food Preparation Machinery* HMSO, London

Health and Safety Executive (1991) *Health and Safety in Motor Vehicle Repair HS(G)67* HMSO, London

Royal Society for the Prevention of Accidents (1972) *Health and Safety in Offices and Shops* RoSPA, Birmingham

Royal Society for the Prevention of Accidents (1976) *Catering Care* RoSPA, Birmingham

Conclusion to Part 4

Safety technology covers an extremely wide field, and a broad under-standing of many engineering disciplines is essential in order to comply with the legal requirements covering machinery, the safe use of electricity, fire protection, mechanical handling and construction activities.

There are still far too many accidents associated with unguarded machinery. Many of these accidents have fatal consequences or can result in people being maimed for life. Application of machinery safety principles detailed in BS 5304 'Safeguarding of machinery' is, therefore, essential if future accidents associated with machinery operation are to be avoided.

The costs to the nation of fire incidents, including those associated with deaths, property damage and lost production, have been substantial in the last decade. Organisations need to examine their fire protection procedures on a regular basis, including enlisting the assistance of the fire authorities wherever guidance is needed. It only needs one major fire to put a company out of business.

Over 1,500 people are killed or injured at work each year as a result of unsafe practices when using electricity or unsafe electrical installations. All staff should be trained in the principles of electrical safety and in safe working practices when using same.

Increasing attention is being paid by the enforcement agencies to the need for a safe working environment, including the elimination of the traditional 'sweat shops' associated with a number of industries. Environmental stress, associated with poor levels of temperature and humidity control, lighting and ventilation, is common in many workplaces. In many industries, high noise levels are accepted by management and workers as an intrinsic feature of that industry, and claims for occupational deafness are a standard business cost. This state of affairs cannot be tolerated and companies should place much greater emphasis on the need to reduce or eliminate all forms of environmental stress. The benefits, as with other areas of health and safety improvement, can be substantial, including reduced absenteeism, reduced health-related claims and improved operator performance.

Index